Lecture Notes in Artificial Intelligence 7828

Subseries of Lecture Notes in Computer Science

W0230490

Manuel Graña   Carlos Toro
Robert J. Howlett   Lakhmi C. Jain (Eds.)

Knowledge Engineering, Machine Learning and Lattice Computing with Applications

16th International Conference, KES 2012
San Sebastian, Spain, September 10-12, 2012
Revised Selected Papers

 Springer

Series Editors

Randy Goebel, University of Alberta, Edmonton, Canada
Jörg Siekmann, University of Saarland, Saarbrücken, Germany
Wolfgang Wahlster, DFKI and University of Saarland, Saarbrücken, Germany

Volume Editors

Manuel Graña
University of the Basque Country
Manuel Lardizabal 1, 20018 San Sebastian, Spain
E-mail: manuel.grana@ehu.es

Carlos Toro
Vicomtech-IK4
Paseo Mijeletegui, 20009 San Sebastian, Spain
E-mail: ctoro@vicomtech.org

Robert J. Howlett
KES International
P.O. Box 2115, Shoreham-by-Sea, BN43 9AF, UK
E-mail: rjhowlett@kesinternational.org

Lakhmi C. Jain
University of Canberra, ACT 2601, Australia
E-mail: lakhmi.jain@unisa.edu.au

ISSN 0302-9743   e-ISSN 1611-3349
ISBN 978-3-642-37342-8   e-ISBN 978-3-642-37343-5
DOI 10.1007/978-3-642-37343-5
Springer Heidelberg Dordrecht London New York

Library of Congress Control Number: 2013933721

CR Subject Classification (1998): I.2, H.4, H.2.8, H.3, I.5, I.4

LNCS Sublibrary: SL 7 – Artificial Intelligence

Typesetting: Camera-ready by author, data conversion by Scientific Publishing Services, Chennai, India

Printed on acid-free paper

Springer is part of Springer Science+Business Media (www.springer.com)

Preface

Knowledge engineering and computational intelligence for information processing are a pervasive phenomenon in our current digital civilization. Huge information resources processed by intelligent systems in everyday applications are expected by the average user with the outmost naturalness. Digital news, socialization of relations, and enhancements derived from the handling of expert decisions are but a few examples of everyday applications. Nevertheless, research in these areas is far from stagnant or having reached the principal goals in its agenda.

Ontologies play a major role in the development of knowledge engineering in various domains, from the Semantic Web down to the design of specific decision-support systems. They are used for the specification of natural language semantics, information modeling and retrieval in querying systems, geographical information systems, and medical information systems, with the list growing continuously. Ontologies allow for easy modeling of heterogeneous information, flexible reasoning for the derivation of consequents or the search of query answers, specification of a priori knowledge, and increasing accumulation of new facts and relations, i.e., reflexive ontologies. Therefore, they are becoming key components of adaptable information processing systems. Classic problems such as ontology matching or instantiation have new and more complex formulations and solutions, involving a mixture of underlying technologies. A good representation of works currently done in this realm are contained in this collection of papers.

Much of modern machine learning has become a branch of statistics and probabilistic system modeling, establishing a sound methodology to assess the value of the systems strongly anchored in statistics and traditional experimental science. Besides, approaches based on nature-inspired computing, such as artificial neural networks, have a broad application and are the subject of active research. They are represented in this collection of papers by several interesting applications.

A very specific new branch of developments is that of lattice computing, gathering works under a simple heading "use lattice operators as the underlying algebra for computational designs." A traditional area of research that falls in this category is mathematical morphology as applied to image processing, where image operators are designed on the basis of maximum and minimum operations; a long track of successful applications support the idea that this approach could be fruitful in the framework of intelligent system design. Works on innovative logical approaches and formal concept analysis can be grouped in this family of algorithms.

For more than 15 years, KES International and its annually organized events have served as a platform for sharing the latest developments in intelligent systems. Organized by the Computational Intelligence Group of the University of the Basque Country and the computer graphics leading institute Vicomtech-IK4, the 16th Annual KES conference was held in the beautiful city of San Sebastian in the north of Spain http://kes2012.kesinternational.org/index.php. This book contains the revised best papers from the general track of this conference. Of over 130 papers received for this general track, only 20 were selected, which represents the most stringent selection standard for conference papers today.

September 2012

Manuel Graña
Carlos Toro
Robert J. Howlett
Lakhmi C. Jain

Table of Contents

Bioinspired and Machine Learning Methods

Machine Learning Applications

Semantics and Ontology Based Techniques

Lattice Computing and Games

Investigation of Random Subspace and Random Forest Regression Models Using Data with Injected Noise

Tadeusz Lasota[1], Zbigniew Telec[2], Bogdan Trawiński[2], and Grzegorz Trawiński[3]

[1] Wrocław University of Environmental and Life Sciences, Dept. of Spatial Management
ul. Norwida 25/27, 50-375 Wrocław, Poland
[2] Wrocław University of Technology, Institute of Informatics,
Wybrzeże Wyspiańskiego 27, 50-370 Wrocław, Poland
[3] Wrocław University of Technology, Faculty of Electronics,
Wybrzeże S. Wyspiańskiego 27, 50-370 Wrocław, Poland
tadeusz.lasota@up.wroc.pl, grzegorz.trawinsky@gmail.com,
{zbigniew.telec,bogdan.trawinski}@pwr.wroc.pl

Abstract. The ensemble machine learning methods incorporating random subspace and random forest employing genetic fuzzy rule-based systems as base learning algorithms were developed in Matlab environment. The methods were applied to the real-world regression problem of predicting the prices of residential premises based on historical data of sales/purchase transactions. The accuracy of ensembles generated by the proposed methods was compared with bagging, repeated holdout, and repeated cross-validation models. The tests were made for four levels of noise injected into the benchmark datasets. The analysis of the results was performed using statistical methodology including nonparametric tests followed by post-hoc procedures designed especially for multiple N×N comparisons.

Keywords: genetic fuzzy systems, random subspaces, random forest, bagging, repeated holdout, cross-validation, property valuation, noised data.

1 Introduction

Ensemble machine learning models have been focussing the attention of many researchers due to its ability to reduce bias and/or variance compared with single models. The ensemble learning methods combine the output of machine learning algorithms to obtain better prediction accuracy in the case of regression problems or lower error rates in classification. The individual classifier or regressor must provide different patterns of generalization, thus the diversity plays an important role in the training process. Bagging [2], boosting [26], and stacking [28] belong to the most popular approaches. In this paper we focus on bagging family of methods. Bagging, which stands for bootstrap aggregating, devised by Breiman [2] is one of the most intuitive and simplest ensemble algorithms providing good performance. Diversity of learners is obtained by using bootstrapped replicas of the training data. That is, different training data subsets are randomly drawn with replacement from the original training set. So obtained training data subsets, called also bags, are used then to train

M. Graña et al. (Eds.): KES 2012, LNAI 7828, pp. 1–10, 2013.
© Springer-Verlag Berlin Heidelberg 2013

different classification and regression models. Theoretical analyses and experimental results proved benefits of bagging, especially in terms of stability improvement and variance reduction of learners for both classification and regression problems [5], [9].

Another approach to ensemble learning is called the random subspaces, also known as attribute bagging [4]. This approach seeks learners diversity in feature space subsampling. All component models are built with the same training data, but each takes into account randomly chosen subset of features bringing diversity to ensemble. For the most part, feature count is fixed at the same level for all committee components. The method is aimed to increase generalization accuracies of decision tree-based classifiers without loss of accuracy on training data. Ho showed that random subspaces can outperform bagging and in some cases even boosting [13]. While other methods are affected by the curse of dimensionality, random subspace technique can actually benefit out of it.

Both bagging and random subspaces were devised to increase classifier or regressor accuracy, but each of them treats the problem from different point of view. Bagging provides diversity by operating on training set instances, whereas random subspaces try to find diversity in feature space subsampling. Breiman [3] developed a method called random forest which merges these two approaches. Random forest uses bootstrap selection for supplying individual learner with training data and limits feature space by random selection. Some recent studies have been focused on hybrid approaches combining random forests with other learning algorithms [11], [16].

We have been conducting intensive study to select appropriate machine learning methods which would be useful for developing an automated system to aid in real estate appraisal devoted to information centres maintaining cadastral systems in Poland. So far, we have investigated several methods to construct regression models to assist with real estate appraisal: evolutionary fuzzy systems, neural networks, decision trees, and statistical algorithms using MATLAB, KEEL, RapidMiner, and WEKA data mining systems [12], [17], [18]. A good performance revealed evolving fuzzy models applied to cadastral data [20], [23]. We studied also ensemble models created applying various weak learners and resampling techniques [15], [21], [22].

The first goal of the investigation presented in this paper was to compare empirically ensemble machine learning methods incorporating random subspace and random forest with classic multi-model techniques such as bagging, repeated holdout, and repeated cross-validation employing genetic fuzzy systems (GFS) as base learners. The algorithms were applied to real-world regression problem of predicting the prices of residential premises, based on historical data of sales/purchase transactions obtained from a cadastral system. The second goal was to examine the performance of the ensemble methods dealing with noisy training data. The susceptibility to noised data can be an important criterion for choosing appropriate machine learning methods to our automated valuation system. The impact of noised data on the performance of machine learning algorithms has been explored in several works, e.g. [1], [14], [24], [25].

2   Methods Used and Experimental Setup

The investigation was conducted with our experimental system implemented in Matlab environment using Fuzzy Logic, Global Optimization, Neural Network, and Statistics toolboxes. The system was designed to carry out research into machine learning algorithms using various resampling methods and constructing and evaluating ensemble models for regression problems.

Real-world dataset used in experiments was drawn from an unrefined dataset containing above 50 000 records referring to residential premises transactions accomplished in one Polish big city with the population of 640 000 within 11 years from 1998 to 2008. The final dataset counted the 5213 samples. Five following attributes were pointed out as main price drivers by professional appraisers: usable area of a flat (*Area*), age of a building construction (*Age*), number of storeys in the building (Storeys), number of rooms in the flat including a kitchen (*Rooms*), the distance of the building from the city centre (*Centre*), in turn, price of premises (*Price*) was the output variable. For random subspace and random forest approaches four more features were employed: floor on which a flat is located (*Floor*), geodetic coordinates of a building (*Xc* and *Yc*), and its distance from the nearest shopping center (*Shopping*).

Due to the fact that the prices of premises change in the course of time, the whole 11-year dataset cannot be used to create data-driven models. In order to obtain comparable prices it was split into 20 subsets covering individual half-years. Then the prices of premises were updated according to the trends of the value changes over 11 years. Starting from the beginning of 1998 the prices were updated for the last day of subsequent half-years using the trends modelled by polynomials of degree three. We might assume that half-year datasets differed from each-other and might constitute different observation points to compare the accuracy of ensemble models in our study and carry out statistical tests. The sizes of half-year datasets are given in Table 1.

Table 1. Number of instances in half-year datasets

1998-2	1999-1	1999-2	2000-1	2000-2	2001-1	2001-2	2002-1	2002-2	2003-1
202	213	264	162	167	228	235	267	263	267
2003-2	2004-1	2004-2	2005-1	2005-2	2006-1	2006-2	2007-1	2007-2	2008-1
386	278	268	244	336	300	377	289	286	181

Table 2. Parameters of GFS used in experiments

Fuzzy system	Genetic Algorithm
Type of fuzzy system: Mamdani	Chromosome: rule base and mf, real-coded
No. of input variables: 5	Population size: 100
Type of membership functions (mf): triangular	Fitness function: MSE
No. of input mf: 3	Selection function: tournament
No. of output mf: 5	Tournament size: 4
No. of rules: 15	Elite count: 2
AND operator: prod	Crossover fraction: 0.8
Implication operator: prod	Crossover function: two point
Aggregation operator: probor	Mutation function: custom
Defuzzyfication method: centroid	No. of generations: 100

As a performance function the root mean square error (RMSE) was used, and as aggregation functions of ensembles arithmetic averages were employed. Each input and output attribute in individual dataset was normalized using the min-max approach. The parameters of the architecture of fuzzy systems as well as genetic algorithms are listed in Table 2. Similar designs are described in [6], [7], [17].

Following methods were applied in the experiments, the numbers in brackets denote the number of input features:

CV(5) – Repeated cross-validation: 10-fold cv repeated five times to obtain 50 pairs of training and test sets, 5 input features pointed out by the experts,

BA(5) – 0.632 Bagging: bootstrap drawing of 100% instances with replacement (Boot), test set – out of bag (OoB), the accuracy calculated as RMSE(BA) = 0.632 x RMSE(OoB) + 0.368 x RMSE(Boot), repeated 50 times,

RH(5) – Repeated holdout: dataset was randomly split into training set of 70% and test set of 30% instances, repeated 50 times, 5 input features,

RS(5of9) – Random subspaces: 5 input features were randomly drawn out of 9, then dataset was randomly split into training set of 70% and test set of 30% instances, repeated 50 times,

RF(5of9) – Random forest: 5 input features were randomly drawn out of 9, then bootstrap drawing of 100% instances with replacement (Boot), test set – out of bag (OoB), the accuracy calculated as RMSE(BA) = 0.632 x RMSE(OoB) + 0.368 x RMSE(Boot), repeated 50 times.

We examined also the impact of noised data on the performance of the aforementioned ensemble methods. Each run of experiment was repeated four times: firstly each output value (price) in training datasets remained unchanged. Next we replaced the prices in 5%, 10%, and 20% of randomly selected training instances with noised values. The noised values were randomly drawn from the bracket [Q1- 1.5 x IQR, Q3+1.5 x IQR], where Q1 and Q2 denote first and third quartile, respectively, and IQR stands for the interquartile range. Schema illustrating the injection of noise is shown in Fig. 1.

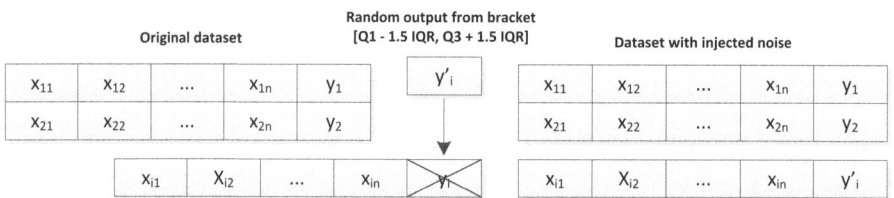

Fig. 1. Schema illustrating the injection of noise

The idea behind statistical methods applied to analyse the results of experiments was as follows. A few articles on the use of statistical tests in machine learning for comparisons of many algorithms over multiple datasets have been published in the last decade [8], [10], [27]. Their authors argue that the commonly used paired tests i.e. parametric t-test and its nonparametric alternative Wilcoxon signed rank tests are not appropriate for multiple comparisons due to the so called family-wise error.

The proposed routine starts with the nonparametric Friedman test, which detect the presence of differences among all algorithms compared. After the null-hypotheses have been rejected the post-hoc procedures should be applied in order to point out the particular pairs of algorithms which produce differences. For $N{\times}N$ comparisons nonparametric Nemenyi's, Holm's, Shaffer's, and Bergamnn-Hommel's procedures are recommended. In the present work we employed the Schaffer's post-hoc procedure due to its relative good power and low computational complexity.

3   Results of Experiments

The accuracy of *BA(5)*, *CV(5)*, *RH(5)*, *RF(5od9)*, and *RS(5of9)* models created using genetic fuzzy systems (GFS) in terms of RMSE for non-noised data and data with the levels of 5%, 10%, and 20% injected noise is shown in Figures 2-5 respectively. In the charts it is clearly seen that *BA(5)* ensembles reveal the best performance, whereas the biggest values of RMSE provide the *RF(5od9)* and *RS(5of9)* models for all levels of noised data. Nonparametric statistical procedures confirm this observation.

Statistical analysis of the results of experiments was performed using a software available on the web page of Research Group "Soft Computing and Intelligent Information Systems" at the University of Granada (http://sci2s.ugr.es/sicidm). In the paper we present the selected output produced by this JAVA software package comprising Friedman tests as well as post-hoc multiple comparison procedures.The Friedman test performed in respect of RMSE values of all models built over 20 half-year datasets showed that there are significant differences between some models. Average ranks of individual models are shown the lower rank value the better model. For all levels of noise the rankings are the same in Table 3, where *BA(5)* reveals the best performance, next are *CV(5)* and *RH(5)*, whereas *RF(5of9)* and *RS(5of9)* are in the last place.

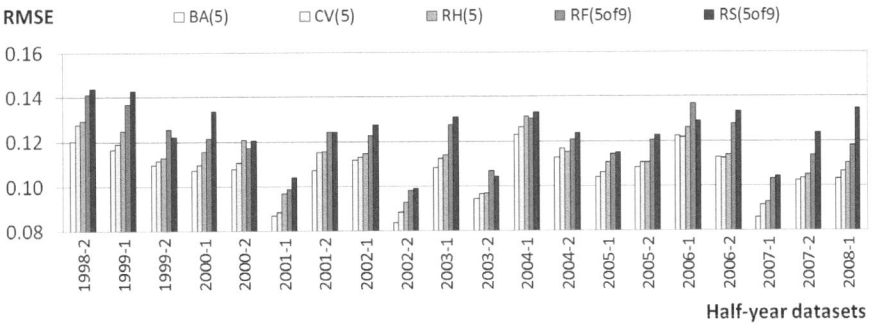

Fig. 2. Performance of ensembles for *non-noised* data

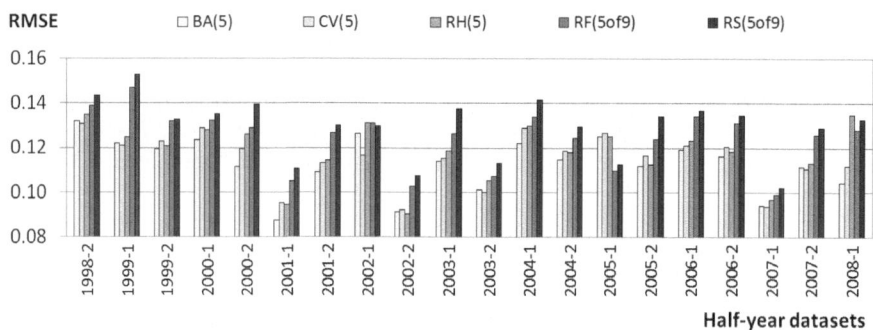

Fig. 3. Performance of ensembles for *5% noised* data

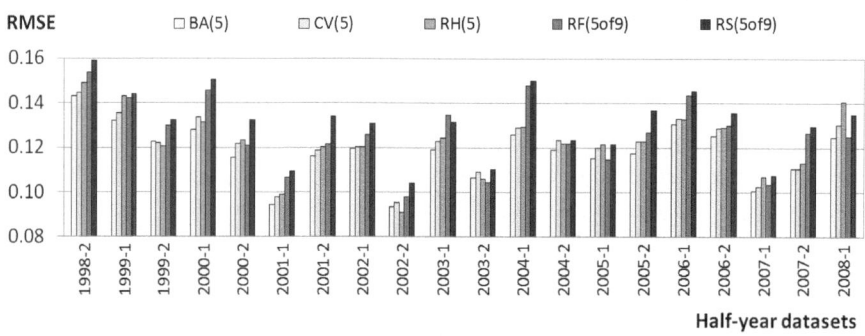

Fig. 4. Performance of ensembles for *10% noised* data

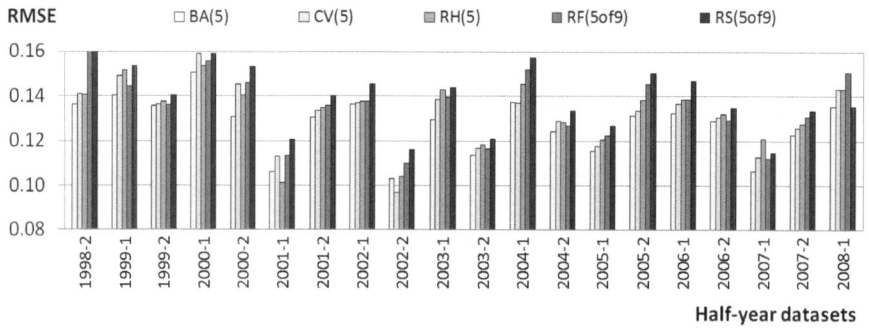

Fig. 5. Performance of ensembles for *20% noised* data

Table 3. Average rank positions of ensembles for different levels of injected noise determined during Friedman test

Rank	1st	2nd	3rd	4th	5th
Method	BA(5)	CV(5)	RH(5)	RF(5of9)	RS(5of9)
0%	1.10	1.95	3.10	4.10	4.75
5%	1.50	2.20	2.80	3.80	4.70
10%	1.35	2.50	2.95	3.35	4.85
20%	1.18	2.55	3.25	3.25	4.78

Table 4. Adjusted p-values produced by Schaffer's post-hoc procedure for $N \times N$ comparisons for all 10 hypotheses for each level of noise. Rejected hypotheses are marked with italics.

Alg. vs Alg.	0%	5%	10%	20%
BA(5) vs RS(5of9)	*2.88E-12*	*1.55E-09*	*2.56E-11*	*6.02E-12*
CV(5) vs RS(5of9)	*0.000000*	*0.000003*	*0.000016*	*0.000052*
BA(5) vs RF(5of9)	*0.000000*	*0.000025*	*0.000380*	*0.000199*
RH(5) vs RS(5of9)	*0.003867*	*0.000868*	*0.000868*	*0.013730*
BA(5) vs RH(5)	*0.000380*	*0.037290*	*0.008246*	*0.000199*
CV(5) vs RF(5of9)	*0.000102*	*0.008246*	0.267393	0.484540
RF(5of9) vs RS(5of9)	0.193601	0.215582	*0.010799*	*0.013730*
BA(5) vs CV(5)	0.178262	0.323027	0.085793	*0.023838*
CV(5) vs RH(5)	0.085793	0.323027	0.736241	0.484540
RH(5) vs RF(5of9)	0.136501	0.182001	0.736241	1.000000

Table 5. Percentage loss of performance for data with 20% noise vs non-noised data [19]

Dataset	CV(5)	BA(5)	RH(5)	RS(5of9)	RF(5of9)
1998-2	10.4%	13.2%	8.9%	16.9%	17.6%
1999-1	25.3%	20.4%	21.4%	7.1%	5.8%
1999-2	22.3%	23.7%	22.0%	15.0%	8.3%
2000-1	44.9%	40.7%	32.7%	18.9%	28.2%
2000-2	31.2%	21.3%	16.0%	27.0%	24.7%
2001-1	27.5%	22.2%	4.4%	15.9%	14.6%
2001-2	15.8%	21.6%	16.6%	12.8%	9.3%
2002-1	21.3%	21.8%	20.0%	14.0%	11.9%
2002-2	9.4%	22.6%	11.8%	17.0%	12.1%
2003-1	23.1%	19.3%	25.4%	9.8%	9.4%
2003-2	20.7%	20.2%	22.1%	15.6%	8.9%
2004-1	8.4%	11.7%	10.8%	18.2%	16.7%
2004-2	10.5%	10.2%	11.2%	7.9%	4.8%
2005-1	10.6%	10.9%	8.9%	10.4%	7.0%
2005-2	20.7%	21.1%	24.8%	22.4%	20.6%
2006-1	12.1%	8.2%	9.8%	13.8%	1.1%
2006-2	16.2%	14.4%	15.5%	1.0%	1.0%
2007-1	22.9%	23.8%	29.8%	9.8%	8.4%
2007-2	21.6%	19.7%	21.5%	8.1%	14.8%
2008-1	34.3%	31.2%	29.9%	0.6%	27.5%
Med.	21.0%	20.7%	18.3%	13.9%	10.7%
Avg	20.5%	19.9%	18.2%	13.1%	12.6%

In Table 4 adjusted p-values for Shaffer's post-hoc procedure for $N \times N$ comparisons are shown for all possible pairs of models. Significant differences resulting in the rejection of the null hypotheses at the significance level 0.05 were marked with italics. Following main observations could be done: BA(5) outperforms any other method except for CV(5) for all levels of noise. The performance of *CV(5)* and *RH(5)* ensembles as well as RH(5) and RF(5of9) is statistically equivalent. R(5of9) revealed significantly worse performance than any other ensemble apart form 0% and 5% noise levels where it is statistically equivalent to RF(5of9).

In our previous work we explored the susceptibility to noise of individual ensemble methods [19]. The most important observation was that in each case the average loss of accuracy for RS(5of9) and RF(5of9) was lower than for the ensembles built over datasets with five features pointed out by the experts. It is clearly seen in Table 5 presenting percentage loss of performance for data with 20% noise versus non-noised data.

Injecting subsequent levels of noise results in worse and worse accuracy of all ensemble methods considered. The Friedman test performed in respect of RMSE values of all ensembles built over 20 half-year datasets indicated significant differences between models. Average ranks of individual methods are shown in Table 6, where the lower rank value the better model. For each method the rankings are the same: ensembles built with non-noised data outperform the others; models with lower levels of noise reveal better accuracy than the ones with more noise. Adjusted p-values produced by Schaffer's post-hoc procedure for N×N comparisons of noise impact on individual method accuracy are placed in Table 7. Statistically significant differences take place between each pair of ensembles with different amount of noise except for 5% and 10% of noise which result in statistically equivalent models.

Table 6. Average rank positions of ensembles for individual methods determined during Friedman test

Rank	1st	2nd	3rd	4th
Noise	0%	5%	10%	20%
CV(5)	1.10	2.15	2.90	3.85
BA5)	1.10	2.10	2.90	3.90
RH(5)	1.40	2.00	2.75	3.85
RS(5of9)	1.25	2.20	2.60	3.95
RF(5of9)	1.40	2.35	2.45	3.80

Table 7. Adjusted p-values produced by Schaffer's post-hoc procedure for $N \times N$ comparisons of noise impact on individual method accuracy. Rejected hypotheses are marked with italics.

noise vs noise	BA(5)	CV(5)	RH(5)	RF(5of9)	RS(5of9)
0% vs 20%	*4.17E-11*	*9.76E-11*	*1.17E-08*	*2.48E-08*	*2.25E-10*
5% vs 20%	*0.000031*	*0.000094*	*0.000018*	*0.001148*	*0.000054*
0% vs 10%	*0.000031*	*0.000031*	*0.002831*	*0.030337*	*0.002831*
10% vs 20%	*0.042918*	*0.039929*	*0.021152*	*0.002831*	*0.002831*
0% vs 5%	*0.042918*	*0.030337*	0.141645	*0.039929*	*0.039929*
5% vs 10%	0.050044	0.066193	0.132385	0.806496	0.327187

4   Conclusions and Future Work

A series of experiments aimed to compare ensemble machine learning methods encompassing random subspace and random forest with classic multi-model techniques such as bagging, repeated holdout, and repeated cross-validation was conducted. The ensemble models were created using genetic fuzzy systems over real-world data taken from a cadastral system. Moreover, the susceptibility to noise of all five ensemble methods was examined. The noise was injected to training datasets by replacing the output values by the numbers randomly drawn from the range of values excluding outliers.

The overall results of our investigation were as follows. Ensembles built using fixed number of features selected by the experts outperform the ones based on features chosen randomly from the whole set of available features. Thus, our research did not confirm the superiority of ensemble methods where the diversity of component models is achieved by manipulating of features. However, the latter seem to be more resistant to noised data. Their performance worsen to a less extent than in the case of models created with the fixed number of features.

We intend to continue our research into resilience to noise regression algorithms employing other machine learning techniques such as neural networks, support vector regression, and decision trees. We also plan to inject noise not only to output values but also to input variables using different probability distributions of noise.

Acknowledgments. This paper was partially supported by the Polish National Science Centre under grant no. N N516 483840.

References

1. Atla, A., Tada, R., Sheng, V., Singireddy, N.: Sensitivity of different machine learning algorithms to noise. Journal of Computing Sciences in Colleges 26(5), 96–103 (2011)
2. Breiman, L.: Bagging Predictors. Machine Learning 24(2), 123–140 (1996)
3. Breiman, L.: Random Forests. Machine Learning 45(1), 5–32 (2001)
4. Bryll, R.: Attribute bagging: improving accuracy of classifier ensembles by using random feature subsets. Pattern Recognition 20(6), 1291–1302 (2003)
5. Bühlmann, P., Yu, B.: Analyzing bagging. Annals of Statistics 30, 927–961 (2002)
6. Cordón, O., Gomide, F., Herrera, F., Hoffmann, F., Magdalena, L.: Ten years of genetic fuzzy systems: current framework and new trends. Fuzzy Sets and Systems 141, 5–31 (2004)
7. Cordón, O., Herrera, F.: A Two-Stage Evolutionary Process for Designing TSK Fuzzy Rule-Based Systems. IEEE Tr. on Sys., Man, and Cyb.-Part B 29(6), 703–715 (1999)
8. Demšar, J.: Statistical comparisons of classifiers over multiple data sets. Journal of Machine Learning Research 7, 1–30 (2006)
9. Fumera, G., Roli, F., Serrau, A.: A theoretical analysis of bagging as a linear combination of classifiers. IEEE Transactions on Pattern Analysis and Machine Intelligence 30(7), 1293–1299 (2008)
10. García, S., Herrera, F.: An Extension on "Statistical Comparisons of Classifiers over Multiple Data Sets" for all Pairwise Comparisons. Journal of Machine Learning Research 9, 2677–2694 (2008)

11. Gashler, M., Giraud-Carrier, C., Martinez, T.: Decision Tree Ensemble: Small Heterogeneous Is Better Than Large Homogeneous. In: 2008 Seventh International Conference on Machine Learning and Applications, ICMLA 2008, pp. 900–905 (2008)
12. Graczyk, M., Lasota, T., Trawiński, B.: Comparative analysis of premises valuation models using keel, rapidminer, and weka. In: Nguyen, N.T., Kowalczyk, R., Chen, S.-M. (eds.) ICCCI 2009. LNCS (LNAI), vol. 5796, pp. 800–812. Springer, Heidelberg (2009)
13. Ho, T.K.: The Random Subspace Method for Constructing Decision Forests. IEEE Transactions on Pattern Analysis and Machine Intelligence 20(8), 832–844 (1998)
14. Kalapanidas, E., Avouris, N., Craciun, M., Neagu, D.: Machine Learning Algorithms: A study on noise sensitivity. In: Manolopoulos, Y., Spirakis, P. (eds.) Proc. 1st Balcan Conference in Informatics 2003, Thessaloniki, pp. 356–365 (November 2003)
15. Kempa, O., Lasota, T., Telec, Z., Trawiński, B.: Investigation of bagging ensembles of genetic neural networks and fuzzy systems for real estate appraisal. In: Nguyen, N.T., Kim, C.-G., Janiak, A. (eds.) ACIIDS 2011, Part II. LNCS, vol. 6592, pp. 323–332. Springer, Heidelberg (2011)
16. Kotsiantis, S.: Combining bagging, boosting, rotation forest and random subspace methods. Artificial Intelligence Review 35(3), 223–240 (2011)
17. Król, D., Lasota, T., Trawiński, B., Trawiński, K.: Investigation of Evolutionary Optimization Methods of TSK Fuzzy Model for Real Estate Appraisal. International Journal of Hybrid Intelligent Systems 5(3), 111–128 (2008)
18. Lasota, T., Mazurkiewicz, J., Trawiński, B., Trawiński, K.: Comparison of Data Driven Models for the Validation of Residential Premises using KEEL. International Journal of Hybrid Intelligent Systems 7(1), 3–16 (2010)
19. Lasota, T., Telec, Z., Trawiński, B., Trawiński, G.: Evaluation of Random Subspace and Random Forest Regression Models Based on Genetic Fuzzy Systems. In: Graña, M., et al. (eds.) Advances in Knowledge-Based and Intelligent Information and Engineering Systems, pp. 88–97. IOS Press, Amsterdam (2012)
20. Lasota, T., Telec, Z., Trawiński, B., Trawiński, K.: Investigation of the eTS Evolving Fuzzy Systems Applied to Real Estate Appraisal. Journal of Multiple-Valued Logic and Soft Computing 17(2-3), 229–253 (2011)
21. Lasota, T., Telec, Z., Trawiński, G., Trawiński, B.: Empirical comparison of resampling methods using genetic fuzzy systems for a regression problem. In: Yin, H., Wang, W., Rayward-Smith, V. (eds.) IDEAL 2011. LNCS, vol. 6936, pp. 17–24. Springer, Heidelberg (2011)
22. Lasota, T., Telec, Z., Trawiński, G., Trawiński, B.: Empirical comparison of resampling methods using genetic neural networks for a regression problem. In: Corchado, E., Kurzyński, M., Woźniak, M. (eds.) HAIS 2011, Part II. LNCS (LNAI), vol. 6679, pp. 213–220. Springer, Heidelberg (2011)
23. Lughofer, E., Trawiński, B., Trawiński, K., Kempa, O., Lasota, T.: On Employing Fuzzy Modeling Algorithms for the Valuation of Residential Premises. Information Sciences 181, 5123–5142 (2011)
24. Nettleton, D.F., Orriols-Puig, A., Fornells, A.: A study of the effect of different types of noise on the precision of supervised learning techniques. Artificial Intelligence Review 33(4), 275–306 (2010)
25. Opitz, D.W., Maclin, R.F.: Popular Ensemble Methods: An Empirical Study. Journal of Artificial Intelligence Research 11, 169–198 (1999)
26. Schapire, R.E.: The strength of weak learnability. Mach. Learning 5(2), 197–227 (1990)
27. Trawiński, B., Smętek, M., Telec, Z., Lasota, T.: Nonparametric Statistical Analysis for Multiple Comparison of Machine Learning Regression Algorithms. International Journal of Applied Mathematics and Computer Science 22(4), 867–881 (2012)
28. Wolpert, D.H.: Stacked Generalization. Neural Networks 5(2), 241–259 (1992)

A Genetic Algorithm vs. Local Search Methods
for Solving the Orienteering Problem
in Large Networks

Joanna Karbowska-Chilińska and Paweł Zabielski

Bialystok University of Technology, Faculty of Computer Science, Poland

Abstract. The Orienteering problem (OP) can be modelled as a weighted graph with set of vertices where each has a score. The main OP goal is to find a route that maximises the sum of scores, in addition the length of the route not exceeded the given limit. In this paper we present our genetic algorithm (GA) with inserting as well as removing mutation solving the OP. We compare our results with other local search methods such as: the greedy randomised adaptive search procedure (GRASP) (in addition with path relinking (PR)) and the guided local search method (GLS). The computer experiments have been conducted on the large transport network (908 cities in Poland). They indicate that our algorithm gives better results and is significantly faster than the mentioned local search methods.

1 Introduction

The Orienteering Problem (OP) is a generalization of well known the traveling salesman problem. The OP is defined as a weighted and complete graph problem on the graph $G = \langle V, E \rangle$, where V and E denote the set of vertices and edges respectively, $|V| = n$. For every vertex a score $p_i \geq 0$ is associated. Additionally each edge has a cost value t_{ij} which could be interpreted as a distance travel, the time or the cost of a travel between vertex i and j. The objective is to find a path from the given starting point s to the destination point e that maximises the total profit and each vertex is visited only once. In addition, the total cost of the edges on this path should be limited by the given t_{max}. It could be interpreted as not exceeded total budget of the tour.

As can be noted the OP is perfect to model the problem of the planning tourist routes. Usually tourists visiting any region or city are not able to visit all attractions because they are constrained not only by the time of the trip but also by the money. Using the above stated solution, they would stand a chance of visiting the most valuable attractions in the presupposed time limit. Several variants of the OP such as the the Orienteering Problem with Time Windows (OPTW) and the Team Orienteering Problem(TOP) are the basis of a software that simplify tour planning [VSVO11], [SV10]. In the OPTW [VSVO9a] for each location the time interval is assigned, in which the location could be visited. The solution of the TOP [VSVO9] generates a set of m routes that have to maximise the total score of the visited points and each of routes satisfied the same constrain on the length. In sophisticated systems of e-tourism another extension

M. Graña et al. (Eds.): KES 2012, LNAI 7828, pp. 11–20, 2013.

of the OP is applied - the Time Dependent Team Orienteering Problem with Time Windows (TDTOPTW) [GVSL10] that allows planning tourist trips and integrates public transport as well as the movements between points of interests (POI) on foot. In the TDTOPTW travel time between POI is not fixed (in the contrary to the OP, TOP, OPTW) because of the various type of public transport, delay in service or traffic jam.

The OP is NP-hard [GLV87], so exact algorithms such as the branch-and-bound as well as the branch -and- cut methods [FST98], [GLS98] are very time consuming (instances up to 500 vertices have been solved). In the practice the following heuristic solutions are usually used instead: genetic algorithms [TS00], local search methods [CGW96], [VSVO9], [CM11], the tabu search [TMO5], the variable neighbourhood search [AHS07], the ant colony optimization approach [SS10].

This article presents our genetic algorithm (GA), with the special mutation operator solving the orienteering problem. To obtain optimal solution we have taken into account in the fitness function not only the total profit but also the travel length for the given path.

Our current research has focused on effective meta-heuristics solutions for the OP which yields the high-quality solutions in large networks. It could be noted the meta-heuristic solutions presented in the literature, provide effective solutions in a reasonable time for only short routes [VSVO11], [GVSL10] (e.g. considering one city or one region of the given country where tours are generated) or are tested on Tsiligirides [Ts84], Fischetti [FST98], Chao [CGW96] benchmarks (the number of vertices between 21 and 500). In the future work we would like to apply our solution in the real- life problems modelled by large networks. Therefore, the tests of the GA and the local search methods (GLS, GRASP, GRASPwPR) for comparison has been carried out on the large transport network which contains 908 Polish cities. The maximum length of generated route has not exceeded 3000 km.

The remaining of this paper is structured as follows. Section 2 outlines an overview of well-know heuristic approaches to the OP. Section 3 gives detailed description of the proposed genetic algorithm. To demonstrate the effectiveness of our method, in Section 4 the results of computational experiments carried out on the large network are discussed. Conclusions and further work are drawn in Section 5.

2   Heuristics Approaches

The orienteering problem has been studied since the eighties. Besides exact algorithms mentioned in the introduction, researches propose heuristics to tackle the OP problem.

One of the first method is the S algorithm [Ts84], based on the Monte Carlo method. During the routes construction new vertices are selected on the basis of probability dependent on the Euclidean distance (between the last point on the route and the analysed vertex) and the score of the vertex under consideration. In the D - Algorithm, another Tsiligrides concept, the paths are built up in separate sectors. In both algorithms, the best path is modified by the path improvement algorithm.

Among the other well-known heuristics should be mention: greedy algorithm with the concept of center of gravity [GLV87], [GWL88] as well as the five-step heuristic of Chao et al. [CGW96], that outperforms all above mentioned heuristics. Schilde et. al [SDHK09] proposed the Pareto ant colony optimisation (ACO) algorithm and the multi objective variable neighbourhood search algorithm (MOVNS), both with path relinking method (PR). Their approaches outperform the five - step heuristic of Chao.

The OP is the special case of the TOP (m=1), therefore some heuristics methods solving the TOP could also be used to tackle the OP. Due to a good benchmark results we mention two methods with the local search framework: the guided local search (GLS) as well as the greedy randomised adaptive search procedure (GRASP) in addition with the path relinking (PR).

Vansteenwegen et al. [VSVO11] developed the GLS for the (T)OP, that reducing the chance of trapping in a local optimum. For each local search iteration, they used special penalties e.g. increase of the score of the non-included locations and decrease of the score of the included location. In an initialization step some of the initial routes are generated [CGW96] and the best one is selected. Every local search iteration consists of three phases: 2-opt procedure, insert and replacement not-included locations. If no replacements are possible, the heuristic reaches a local optimum and then the penalty function is applied. Next, replacements are repeated a fixed number of times and the local search iteration of a diversification procedure is applied to explore the whole solution space.

Souffriau et al. [SVVO10] applied the GRASP and the PR approach for the TOP. Campos et al. [CM11] adapted the GRASP and the PR approach to solve the OP. They revealed that these methods give better results on benchmark instances than others. In the GRASP algorithm proposed by them the initial path contains only starting and ending vertex. Next, based on the ratio between greediness and randomness (four methods are possible), vertices are inserted one by one as long as t_{max} is not exceeded. In the next step the local search procedure (exchange and insert phase) try reduce the length of a path and increase its total profit. In GRASPwPR [CM11] set of different solutions is created with GRASP method and for each pair of solutions P and Q a path relinking is performed: the P solution is gradually transformed into the Q, by exchanging elements between P and Q.

Solutions mentioned in the literature were usually tested in small networks (the number of vertices in Tsiligirides [Ts84], Chao et al. [CGW96] and Fischetti et al. [FST98] benchmark instances vary between 21 and 500). Therefore one of goals was to conduct tests for the GA and the selected local search methods in the large transport network and to compare the effectiveness of methods. The experiment results are drawn in Section 4.

3   Overview of the GA

The described method, first presented in [KKOZ12] is an improved version of the algorithm published on [OK11], [PK11]. The GA works on the standard version of the OP assuming that the given graph satisfies the triangle inequality.

In the GA algorithm, after generation of initial population the following procedures: tournament selection, crossover and mutation are executed in a loop. The algorithm stops after n_g generations or no further improvements are identified during fixed number of iterations. The fitness function F estimates the quality of individuals, according to the sum of profits $TotalProfit$ and the length of the tour $TravelLength$. Each vertex can be visited more than once in the route but the $TotalProfit$ is increased when the vertex is visited for the first time. F is equal to $TotalProfit^3/TravelLength$. The output of the GA is a tour with the highest value of F from the final generation.

3.1   Construction

First, the initial population of P_{size} solutions is generated. Tours are encoded into chromosomes as a sequence of vertices (cities). It is the most natural way of adapting the GA for the OP. At the beginning a random vertex v, which is adjacent to vertex s (the start point) is chosen. Next, the value t_{sv} is added to the current travel length. If the current travel length does not exceed $0.5 \cdot t_{max}$, the tour generation is continued. Now we start at vertex v and choose a random vertex u adjacent to v. At every step we remove the previously visited vertex from the set of unvisited vertices. If the current tour length is greater than $0.5 \cdot t_{max}$, the last vertex is rejected and we return to vertex s the same way but in the reversed order. The initial routes are symmetrical with respect to their middle vertex in the tour. However, these symmetries are removed by the algorithm.

3.2   Tournament Selection

In the first step, we select t_{size} random, numerous individuals from the current population, and the best one from the group is copied to the next population. It is important that we choose the best chromosome from the group that is the one with the highest value of $TotalProfit^3/TravelLength$. Next, the whole tournament group is returned to the old population, and finally after P_{size} repetitions of this step a new population is created.

3.3   Crossover

First two random individuals are selected as parents. Afterwards we randomly choose a common gene (crossing point) from both parents (the first and the last genes are not considered). If there are no common genes, crossover does not occur and no changes are applied. Following this, two new individuals are created as a result of exchanging chromosome fragments (from the crossing point to the end of the chromosome) from both parents. If one of the children does not preserve t_{max}

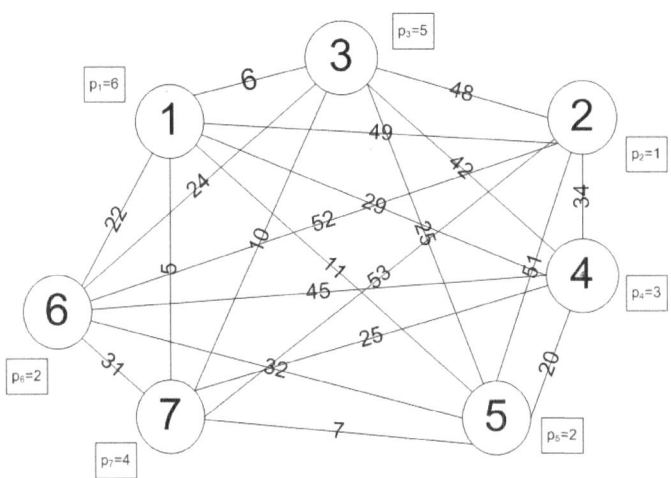

Fig. 1. Weight graph G example - the numbers next to vertices are profits

constraint, the fitter parent from the new population replaces it. If both children do not preserve this constraint, the parents replace them in the new population (no changes applied). In the example presented in fig. 2 (graph in fig.1) two children with travel length not exceeded t_{max} are created. Parents can be symmetrical, however the crossover removes the symmetry from the offspring individuals.

3.4   Mutation

After the selection and crossover phases, the population undergoes the heuristic mutation in the version inserting a new gene or removing an existing one (with probability 0.5). First, a random individual is selected to be mutated. We apply the insertion mutation in which all possibilities of inserting a new available gene u that is not present in the chromosome are considered. The place with the highest value of $p_u^2/TravelLengthIncrease(u)$ ($TravelLengthIncrease(u)$ - increase of the travel length after u is inserted to the chromosome) is chosen. If $TravelLengthIncrease(u)$ is less than 1, we must consider the highest value of p_u^2. The gene u is not available when the route exceeds t_{max} or does not increase the value of currently best fitness function. For example, in fig. 3 there is only one possibility inserting a new vertex (the gen 5 between 1 and 3) into the given route without exceeding t_{max}. In this case the insert mutation could not be performed because it does not cause the increase the current value of the fitness function.

   In the removing mutation, only genes that appear in the chromosome more than once are considered (with the exception of the first and last genes). If there are no such candidates the removing mutation is not performed (no fitness loss occurs). Otherwise, we choose the gene in order to shorten the travel length as much as possible. For example, in fig. 4 we have only one candidate suitable for removal, gene 3. If we remove this gene from the first position, the path is shorten by 20, but if we remove gene 3 from the second position the path is shorten by 11. Thus, the gen from the first position is removed.

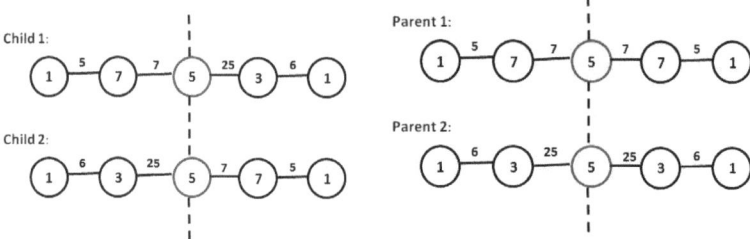

Fig. 2. Crossover example (where $t_{max} = 134$), the crossing point is the vertex 5

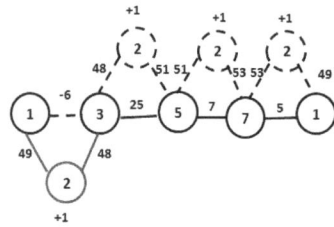

Fig. 3. Insert mutation (where $t_{max} = 134$), where the inclusion of the vertex 5 between the vertex 1 and 3 does not improve fitness function

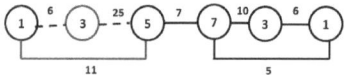

Fig. 4. Removing mutation ($t_{max} = 134$), where the gen 3 (between the gen 1 and 5) is removed

4   Computational Experiment and Results

The our genetic algorithm GA and other heuristics such as: GLS, GRASP and GRASPwPR have been implemented and run on an Intel Core 2 duo 2.8GHz CPU. The experiments have been carried out on the instance presented below.

4.1   Test Instance

The large instance for the OP has been proposed in [KKOZ12] and it has been used by us [Ko12]. The database contains 908 cities in Poland. Each city is described by its longitude, latitude as well as its profit. Profit of each city is equal to the number of inhabitants divided by 1000. The values of profits are between 1 and 1720.

4.2   Experiment Details

The spherical distances between cities has been taken into account. The vertex 1, which corresponds to the capital of Poland, has been established as the starting and the ending point of the route. All tests are conducted on the following values of t_{max}: 500, 1000, 1500, 2000, 2500, 3000 kilometres. The GA parameters are determined experimentally: $P_{size} = 300$, $t_{size} = 5$. The maximal number of the GA generations (n_g) equals to 1000 but every 100 generations the GA checks population and stops if no improvements have been found. The route with the best fitness results from 30 algorithm runs and the execution time all runs are taken into account in the final statistics. In the tests the GLS parameters are the same as in [VSVO9] with the exception of the replace phase. In this step, the maximum number of iterations is determined by tests and it is set to 700 [KKOZ12]. The implementation of the GRASP(wPR) is based on the paper [CM11]. Four different constructive methods (denoted as C1, C2, C3, C4), each of them with four values of randomness (α equals to 0.2, 0.4, 0.6 or 0.8) were tested in the GRASP. The method C2 ($\alpha=0.2$), giving the best profit result in a reasonable short time, was chosen to the final statistic. In the GRASP, 500 paths were constructed and the best one was improved by the local search method. In the GRASPwPR, first the GRASP for 500 constructions was run and next the path relinking to all pair of paths was applied.

4.3   Results

The comparison of the results from the genetic algorithm to those from the GLS as well as the GRASP and the GRASPwPR is outlined in tab.1. The execution time is given in seconds and the percentage gap with our solution is calculated by the following formula: $(GA\_score\text{-}other\_method\_score)/GA\_score$. The last row in tab.1 presents the average values of obtained results of the GA and the other methods. Results reveals that the GA is superior to all methods, for almost all tested cases. The percentage difference between the GA and the GLS scores varies between 2% for $t_{max} = 2500$ and 12.6% for $t_{max} = 1000$. Only in the case $t_{max} = 1500$, the GA gives slightly worse scores (0.8%) than GLS, but the GA is significantly faster. The difference between the GA and the GRASP varies between 0.4% for $t_{max} = 500$ and 8.5% for $t_{max} = 2500$. Results indicate that in the case of large networks the path relinking procedure slightly improves the GRASP method (on average 0.4%) but the execution time of the GRASPwPR is significantly higher than the GRASP. Note that in the presented experiment, the average execution time of the GA is approximately three times lower than the GRASP and two times lower than GLS. In fig.5, examples of the best-generated routes are marked on the Polish map for the GA, the GLS as well as the GRASP method. Usually final paths are comparable in length (for the same value of t_{max}) but they pass through various cities.

Table 1. Comparison of GA, GLS, GRASP, GRASPwPR results and execution time

	GA		GLS		GRASP		GRASPwPR		% gap GA with		
t_{max}	score	time	score	time	score	time	score	time	GLS	GRASP	GRASPwPR
500	3666	2.1	3456	4.0	3652	16.6	3631	27.1	3.27	0.38	0.95
1000	7621	4.2	6659	14.7	7267	21.8	7472	31.1	12.62	4.65	1.96
1500	9671	5.5	9750	19.1	8859	20.9	8862	37.2	-0.81	8.40	8.37
2000	10527	6.9	10236	17.8	10280	24.4	10256	45.0	2.76	2.35	2.57
2500	12280	12.9	12027	22.5	11234	31.6	11400	54.1	2.06	8.52	7.17
3000	13237	17.2	12464	21.5	12595	41.5	12472	67.4	5.85	4.85	5.78
Avg.	9500.3	8.1	9113.7	16.6	8981.2	26.1	9015.5	43.6	4.29	4.86	4.47

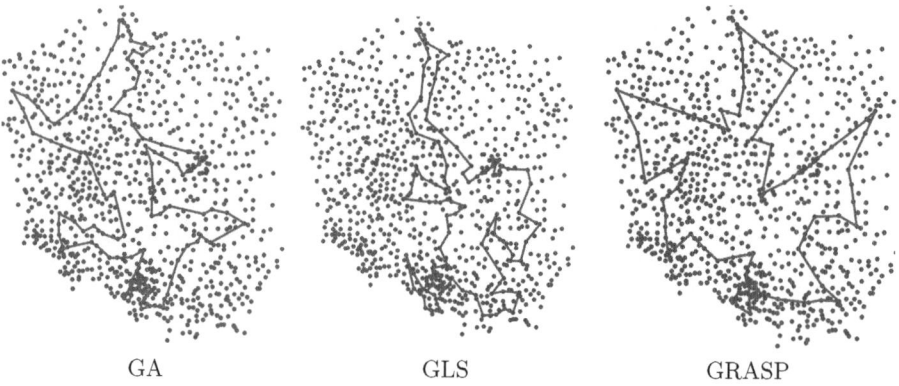

GA   GLS   GRASP

Fig. 5. Examples of routes generated for $T_{max} = 3000$

5   Conclusion and Further Work

The computer experiments present that the GA performs better than comparable local search methods (the GLS, the GRASP and the GRASPwPR). Most of the best known solutions were tested on small networks (between 21 and 500 vertices) [Ts84], [CGW96], [FST98] but in this article all algorithms have been tested on a large network (908 vertices) with six instance of t_{max}. In the real world we can find a lot of examples where the number of vertices is greater than 500 and this is the reason why this element is so important. The GA is better than other efficient local search methods. In addition, the execution time is also better. Future research on the GA should focus on solving other instances of large networks. We will try to change some steps to improve the implemented algorithm. In every evolutionary operator we want to reduce an execution time. After that we will work on the algorithm that will solve the team orienteering problem where m routes are required.

Acknowledgements. The authors gratefully acknowledge support from the Polish Ministry of Science and Higher Education at the Bialystok University of Technology (grant S/WI/1/2011 and S/WI/2/2008).

References

[AHS07]   Archetti, C., Hertz, A., Speranza, M.G.: Metaheuristics for the team orienteering problem. Journal of Heuristics 13, 49–76

[Ba89]   Balas, E.: The prize collecting traveling salesman problem. Networks 19, 797–809 (1989)

[CGW96]   Chao, I.M., Golden, B.L., Wasil, E.A.: A Fast and effective heuristic for the orienteering. European Journal of Operational Research 88, 475–489 (1996)

[CM11]   Campos, V., Marti, R.: Grasp with Path Relinking for the Orienteering Problem. Technical Raport, 1–16 (2011)

[FST98]   Fischetti, M., Salazar, J.J., Toth, P.: Solving the Orienteering Problem through Branch-and-Cut. INFORMS Journal on Computing 10, 133–148 (1998)

[GLS98]   Gendreau, M., Laporte, G., Semet, F.: A branch-and-cut algorithm for the undirected selective traveling salesman problem. Networks 32(4), 263–273 (1998)

[GVSL10] Garcia, A., Arbelaitz, O., Vansteenwegen, P., Souffriau, W., Linaza, M.T.: Hybrid approach for the public transportation time dependent orienteering problem with time windows. In: Corchado, E., Graña Romay, M., Manhaes Savio, A. (eds.) HAIS 2010, Part II. LNCS, vol. 6077, pp. 151–158. Springer, Heidelberg (2010)

[GLV87]   Golden, B., Levy, L., Vohra, R.: The orienteering problem. Naval Research Logistics 34, 307–318 (1987)

[GWL88]   Golden, B., Wang, Q., Liu, L.: A multifaceted heuristic for the orienteering problem. Naval Research Logistics 35, 359–366 (1988)

[KKOZ12] Karbowska-Chilinska, J., Koszelew, J., Ostrowski, K., Zabielski, P.: Genetic algorithm solving orienteering problem in large networks. Frontiers in Artificial Intelligence and Applications 243 (2012)

[Ko12]   Ostrowski, K., http://jolantakoszelew.pl/attachments/File/Poland908cities.txt (last access: November 15, 2012)

[OK11]   Ostrowski, K., Koszelew, J.: The comparision of genetic algorithm which solve Orienteering Problem using complete an incomplete graph. Zeszyty Naukowe, Politechnika Bialostocka, Informatyka 8, 61–77 (2011)

[PK11]   Piwońska, A., Koszelew, J.: A memetic algorithm for a tour planning in the selective travelling salesman problem on a road network. In: Kryszkiewicz, M., Rybinski, H., Skowron, A., Raś, Z.W. (eds.) ISMIS 2011. LNCS (LNAI), vol. 6804, pp. 684–694. Springer, Heidelberg (2011)

[SDHK09]   Schilde, M., Doerner, K., Hartl, R., Kiechle, G.: Metaheuristics for the bi-objective orienteering problem. Swarm Intelligence 3, 179–201 (2009)

[SS10]   Sevkli, Z., Sevilgen, E.: Discrete particle swarm optimization for the orienteering Problem. In: IEEE Congress (2010)

[So10]   Souffriau, W.: Automated Tourist Decision Support, PhD Thesis, Katholieke Universiteit Leuven (2010)

[SV10]   Souffriau, W., Vansteenwegen, P.: Tourist trip planning functionalities: state–of–the–art and future. In: Daniel, F., Facca, F.M. (eds.) ICWE 2010. LNCS, vol. 6385, pp. 474–485. Springer, Heidelberg (2010)

[SVVO10]   Souffriau, W., Vansteenwegen, P., Vanden Berghe, G., Van Oudheusden, D.: A path relinking approach for the team orienteering problem. Computers & Operational Research 37, 1853–1859 (2010)

[TMO5]   Tang, H., Miller-Hooks, E.: A tabu search heuristic for the team orienteering problem. Comput. Oper. Res. 32(6), 1379–1407 (2005)

[Ts84]   Tsiligirides, T.: Heuristic methods applied to orienteering. Journal of the Operational Research Society 35(9), 797–809 (1984)

[TS00]   Tasgetiren, M.F., Smith, A.E.: A genetic algorithm for the orienteering problem. In: Proceedings of the 2000 Congress on Evolutionary Computation, San Diego, vol. 2, pp. 1190–1195 (2000)

[VSVO9]   Vansteenwegen, P., Souffriau, W., Vanden Berghe, G., Van Oudheusden, D.: A guided local search metaheuristic for the team orienteering problem. European Journal of Operational Research 196, 118–127 (2009)

[VSVO9a]   Vansteenwegen, P., Souffriau, W., Vanden Berghe, G., Van Oudheusden, D.: Iterated local search for the team orienteering problem with time windows. Computers O.R. 36, 3281–3290 (2009)

[VSVO11]   Vansteenwegen, P., Souffriau, W., Vanden Berghe, G., Van Oudheusden, D.: The City Trip Planner: An expert system for tourists. Expert Systems with Applications 38(6), 6540–6546 (2011)

[VSVO9b]   Vansteenwegen, P., Souffriau, W., Vanden Berghe, G., Van Oudheusden, D.: Metaheuristics for tourist trip planning. LNEMS, vol. 624, pp. 15–31 (2009)

Dynamic Structure of Volterra-Wiener Filter for Reference Signal Cancellation in Passive Radar*

Pawel Biernacki

Telecom, Acoustic and Computer Science Institute,
Wroclaw University of Technology,
Wyb. Wyspianskiego 27, 50-350 Wroclaw, Poland
pawel.biernacki@pwr.wroc.pl.com

Abstract. In the article a possibility of using the Volterra-Wiener filter for reference signal elimination in passive radar was considered. The recursive nonlinear orthogonal filter algorithms (with low-complexity and dynamic structures) were developed and implemented within Matlab environment. The results of testing with real-life data are comparable with the effects of the NLMS filter algorithm employment.

Keywords: parallel computing, orthogonal filter, multidimensional signal representation.

1 Introduction

The passive radar systems use the FM radio signal as a source to find the flying objects. A critical limitation of such a system is the unwanted interference in the echo channel due to the direct reception of the FM radio signal (reference signal). This unwanted direct signal can be over 90dB greater than the wanted target echo. Thus, in order to detect target echoes, it is necessary to suppress the direct signal by as much as possible.

The two following (complex) algorithms for the reference signal cancellation were employed:

- Decorrelated NLMS (DNLMS) - described in [3],
- Nonlinear Orthogonal Filter (NOF) - described below,

All algorithms were implemented in the MATLAB environment using the fixed-point arithmetic to check a possibility of their implementation within the VHDL environment. Simulations results, for the real-life signals are presented in the Simulation section.

2 Reference Signal Cancellation Using Volterra-Wiener Class Filter

Given a vector $|y>_T$ of samples $\{y_0, \ldots, y_T\}$ of a time-series (a reference signal + an echo signal), observed on a finite time interval $\{0, \ldots, T\}$, the reference

---

* Work sponsored by NC3 NATO Agency.

M. Graña et al. (Eds.): KES 2012, LNAI 7828, pp. 21–30, 2013.
© Springer-Verlag Berlin Heidelberg 2013

signal cancellation problem can be stated geometrically as follows (see Figure 1). The estimate of the desired signal

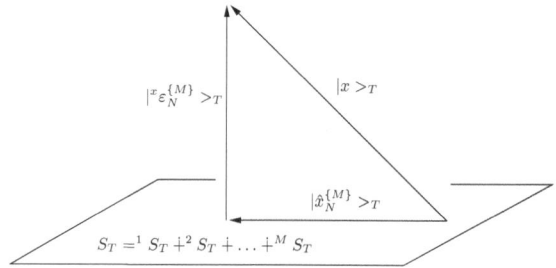

Fig. 1. The estimate $|\hat{x} >_T$ of the desired signal

$$|\hat{x}_N^{\{M\}} >_T \triangleq \mathbf{P}(S_T)|x >_T \tag{1}$$

is the orthogonal projection of the element $|x >_T$ (echo signal) on the space S_T spanned by the following set of the linear and nonlinear observations

$$|Y >_T = [|^1Y >_T \;\; |^2Y >_T \ldots |^MY >_T] \tag{2}$$

where

$$|^mY >_T = [|y_{i_1} \ldots y_{i_m} >_T; \; i_1 = 0, \ldots, N,$$
$$i_2 = i_1, \ldots, N, \ldots, i_m = i_{m-1}, \ldots, N] \tag{3}$$

for $m = 1, \ldots, M$. The orthogonal projection operator on $|Y >_T$ is defined as

$$\mathbf{P}(S_T) \triangleq |Y >_T < Y|Y >_T^{-1} < Y|_T \tag{4}$$

If an ON (generalized; i.e., multidimensional) basis of the space S_T is known, the projection operator on S_T can be decomposed as

$$\mathbf{P}(S_T) = \sum_{j_1=0}^{N} \mathbf{P}(|r_0^{j_1} >_T) + \ldots +$$

$$+ \sum_{j_1=0}^{N} \ldots \sum_{j_M=j_{M-1}}^{N} \mathbf{P}(|r_0^{j_1,\ldots,j_M} >_T) \tag{5}$$

where $\mathbf{P}(|r_0^{j_1,\ldots,j_m} >_T)$ stands for the orthogonal projection operator on the one-dimensional subspace spanned by the element $r_0^{j_1,\ldots,j_m}$, $m = 1, \ldots, M$ of an ON basis of the space S_T. Since

$$\mathbf{P}(|r_0^{j_1,\ldots,j_w} >_T) = |r_0^{j_1,\ldots,j_w} >_T < r_0^{j_1,\ldots,j_w}|_T \tag{6}$$

the orthogonal expansion of the estimate of the desired signal can be written as

$$|\hat{x}_N^{\{M\}} >_T = \mathbf{P}(S_T)|x>_T = \sum_{j_1=0}^{N} |r_0^{j_1} >_T < r_0^{j_1}|x>_T +$$

$$+ \ldots + \sum_{j_1=0}^{N} \cdots \sum_{j_M=j_{M-1}}^{N} |r_0^{j_1,\ldots,j_M} >_T < r_0^{j_1,\ldots,j_M}|x>_T \qquad (7)$$

The estimation error associated with the element $|\hat{x}_N^{\{M\}} >_T$ is then

$$|^x\varepsilon_N^{\{M\}} >_T \triangleq \mathbf{P}(S_T^\perp)|x>_T = |x>_T - |\hat{x}_N^{\{M\}} >_T \perp S_T \qquad (8)$$

The estimate (1) will be called optimal (in the least-squares sense) if the norm

$$\| \, |^x\varepsilon_N^{\{M\}} >_T \, \| = <^x \varepsilon_N^{\{M\}}|^x\varepsilon_N^{\{M\}} >_T^{\frac{1}{2}} \qquad (9)$$

of the estimation error vector (8) is minimized for each $T = 0, 1, 2, \ldots$.

The nonlinear reference signal cancellation problem can, therefore, be solved in the following two steps: a) derivation of a (generalized) ON basis of the estimation space S_T, b) calculation of the orthogonal representation (i.e., the generalized Fourier coefficients) of the vector $|x>_T$ in the orthogonal expansion (7).

To derive the desired ON basis of the estimation space S_T (Step a)), we employ (consult [1][5]) the following

Theorem 1. *The partial orthogonalization step results from the recurrence relations*

$$|e_{i_1,\ldots,i_q}^{j_1,\ldots,j_w} >_T = [|e_{i_1,\ldots,i_q}^{j_1,\ldots,j_w-1} >_T +$$

$$+ |r_{i_1,\ldots,i_q+1}^{j_1,\ldots,j_w} >_T \, \rho_{i_1,\ldots,i_q;T}^{j_1,\ldots,j_w}](1 - (\rho_{i_1,\ldots,i_q;T}^{j_1,\ldots,j_w})^2)^{-\frac{1}{2}} \qquad (10)$$

$$|r_{i_1,\ldots,i_q}^{j_1,\ldots,j_w} >_T = [|e_{i_1,\ldots,i_q}^{j_1,\ldots,j_w-1} >_T \, \rho_{i_1,\ldots,i_q;T}^{j_1,\ldots,j_w} +$$

$$+ |r_{i_1+1,\ldots,i_q+1}^{j_1,\ldots,j_w} >_T](1 - (\rho_{i_1,\ldots,i_q;T}^{j_1,\ldots,j_w})^2)^{-\frac{1}{2}} \qquad (11)$$

where

$$\rho_{i_1,\ldots,i_q;T}^{j_1,\ldots,j_w} = - < e_{i_1,\ldots,i_q}^{j_1,\ldots,j_w-1}|r_{i_1,\ldots,i_q+1}^{j_1,\ldots,j_w} >_T \qquad (12)$$

Proof can be found in [1].

The above relations make it possible to construct an orthogonal decorrelation filter, operating directly on the signal samples, and allowing for real-time implementation of the decorrelation block. Introducing the so-called 'information normalization' [4] for the forward as well as backward prediction error vectors, we obtain the time-update formulae for the multi-dimensional Schur coefficients

$$\rho_{i_1,\ldots,i_q;T}^{j_1,\ldots,j_w} = \rho_{i_1,\ldots,i_q;T-1}^{j_1,\ldots,j_w}(1 - (e_{i_1,\ldots,i_q;T}^{j_1,\ldots,j_w-1})^2)^{\frac{1}{2}} \cdot$$

$$\cdot (1 - (r_{i_1,\ldots,i_q+1;T}^{j_1,\ldots,j_w})^2)^{\frac{1}{2}} - e_{i_1,\ldots,i_q;T}^{j_1,\ldots,j_w-1} r_{i_1,\ldots,i_q+1;T}^{j_1,\ldots,j_w} \qquad (13)$$

as well as for the forward and backward prediction error samples

$$
\mathrm{e}^{j_1,\ldots,j_w}_{i_1,\ldots,i_q;T} = (1 - (\rho^{j_1,\ldots,j_w}_{i_1,\ldots,i_q;T})^2)^{-\frac{1}{2}}(1 - (\mathrm{r}^{j_1,\ldots,j_w}_{i_1,\ldots,i_q+1;T})^2)^{-\frac{1}{2}} \cdot
$$
$$
\cdot(\mathrm{e}^{j_1,\ldots,j_w-1}_{i_1,\ldots,i_q;T} + \rho^{j_1,\ldots,j_w}_{i_1,\ldots,i_q;T}\mathrm{r}^{j_1,\ldots,j_w}_{i_1,\ldots,i_q+1;T}) \tag{14}
$$

$$
\mathrm{r}^{j_1,\ldots,j_w}_{i_1,\ldots,i_q;T} = (1 - (\rho^{j_1,\ldots,j_w}_{i_1,\ldots,i_q;T})^2)^{-\frac{1}{2}}(1 - (\mathrm{e}^{j_1,\ldots,j_w-1}_{i_1,\ldots,i_q;T})^2)^{-\frac{1}{2}} \cdot
$$
$$
\cdot(\mathrm{e}^{j_1,\ldots,j_w-1}_{i_1,\ldots,i_q;T}\rho^{j_1,\ldots,j_w}_{i_1,\ldots,i_q;T} + \mathrm{r}^{j_1,\ldots,j_w}_{i_1,\ldots,i_q+1;T}) \tag{15}
$$

These equations make possible the real-time implementation of each elementary decorrelation section of the system.

The above recurrence relations actually solve the problem of the real-time derivation of the (generalized) ON basis of the estimation space (Step a)).

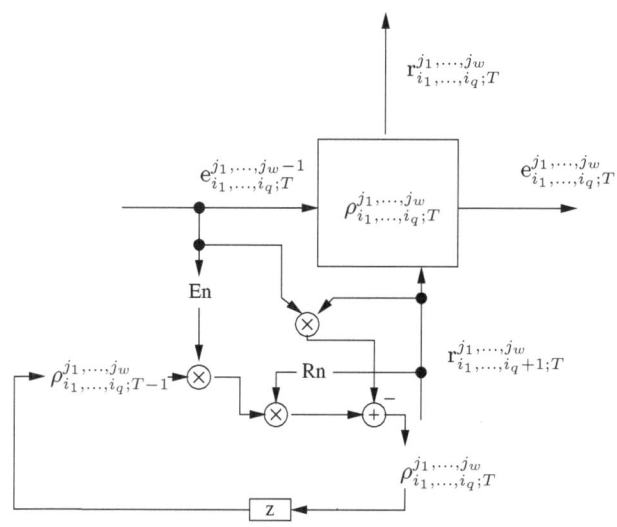

Fig. 2. The elementary section of the decorrelation block

To compute (Step b)) the orthogonal representation of the nonlinear estimate of $|x >_T$ in the orthogonal expansion (7), let us develop the time-update formulae for the generalized Fourier coefficients ${}^x\rho^{j_1,\ldots,j_w}_T$ as well as for the estimates ${}^x\mathrm{e}^{j_1,\ldots,j_w}_T$. Considering

$$
|{}^x\mathrm{e}^{j_1,\ldots,j_w} >_T \triangleq |{}^x\mathrm{e}^{j_1,\ldots,j_w} >_T < \pi|\mathbf{P}(S_T)^\perp|\pi >_T^{-\frac{1}{2}} \tag{16}
$$

and following [3-9], we get

$$
{}^x\rho^{j_1,\ldots,j_w}_T = (1 - (\mathrm{r}^{j_1,\ldots,j_w}_0)^2)^{\frac{1}{2}}(1 - ({}^x\mathrm{e}^{j_1,\ldots,j_w-1}_T)^2)^{\frac{1}{2}} \cdot
$$
$$
\cdot {}^x\rho^{j_1,\ldots,j_w}_{T-1} + {}^x\mathrm{e}^{j_1,\ldots,j_w-1}_T\mathrm{r}^{j_1,\ldots,j_w}_0 \tag{17}
$$

and

$$
{}^x e_T^{j_1,\ldots,j_w} = (1 - (r_0^{j_1,\ldots,j_w})^2)^{-\frac{1}{2}}(1 - {}^x \rho_T^{j_1,\ldots,j_w})^{-\frac{1}{2}} \cdot
$$
$$
\cdot \left[{}^x e_T^{j_1,\ldots,j_w-1} - {}^x \rho_T^{j_1,\ldots,j_w} r_0^{j_1,\ldots,j_w} \right] \tag{18}
$$

The presented relations (17) and (18) make possible a real-time realization of the estimation block of the reference signal cancellation filter.

The diagram of the reference signal cancellation filter is presented in Figure 3. The linear orthogonal reference signal cancellation time-update filter is a special case of the procedure presented above, corresponding to $M = 1$.

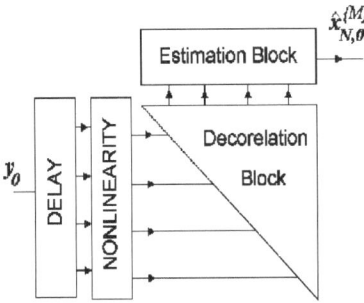

Fig. 3. Nonlinear orthogonal reference signal cancellation filter

3   Filter Parameters Selection Strategy

The aim of increasing degree of filter nonlinearity or its order is a improvement of the echo cancelling quality (minimizing (19))[6].

$$
{}^x R_{N,0}^{\{M\}} = \|x_0\|_\Omega^2 - \sum_{j_1=0}^{N} |{}^x \rho^{j_1}|^2 - \ldots - \sum_{j_1=0}^{N} \ldots \sum_{j_M=j_{M-1}}^{N} |{}^x \rho^{j_1,\ldots,j_M}|^2 \tag{19}
$$

To judge the proper values of N and M the objective measure is needed, which shows the relative improvement of the echo cancelling quality. It should describe a change of the estimation error causes by extension of filter structure (increasing the number of elementary sections in the decorelation block). The following cost function is defined

$$
FK(N_1, M_1; N_2, M_2) \triangleq - \frac{\dfrac{K_{N_2,M_2}^{ns} - K_{N_1,M_1}^{ns}}{K_{N_1,M_1}^{ns}}}{\dfrac{{}^x R_{N_2}^{\{M_2\}} - {}^x R_{N_1}^{\{M_1\}}}{{}^x R_{N_1}^{\{M_1\}}}} \tag{20}
$$

where N_1, N_2 are the filter orders, M_1, M_2 are the filter nonlinearity degrees. The value $K_{N,M}^{ns}$ (21)

$$K_{N,M} = \sum_{m=1}^{M} \frac{(N+m-1)!}{m!(N-1)!} \qquad (21)$$

describes the number of elementary sections in the estimation block for filter of order N and nonlinearity degree of M. ^xR is defined in (19). The equation (20) is interpreted as a relative change of the number of elementary sections in the decorelation block to relative improvement of a echo cancelling quality (change value of the estimation error) when the parameters N, M are changing. Because the denominator is always negative the sign minus is used. This cost function allows to judge efficiency of the selected filter structure and the filter complexity needed to improved the estimate $\hat{x}_0$.

When one or few coefficients $^x\rho$ are added to the estimation it is not necessary to compute the whole ON basis anew, but only its new element. The filter $F_a(N_1, M_1)$ has $L_{N_1,M_1}^{ns} = K_{N_1,M_1}(K_{N_1,M_1}+1)/2$ elements in the decorelation block, where $K_{N,M}$ is described in (21). The value of cost function (20) after adding a new coefficient $^x\rho_{new}$ (it means adding $K_{N_1,M_1}+1$ elementary sections to the decorelation block) is

$$FK(N_1, M_1; N_2, M_2) = \frac{2}{K_{N_1,M_1}} \cdot \frac{^xR_{N_1}^{\{M_1\}}}{|^x\rho_{new}|^2} \qquad (22)$$

Using (22) and considering presented discussion the following strategy for filter structure selection is proposed

Strategy 1. *If K_{N_1,M_1} is the number the coefficients $^x\rho$ for the filter $F_a(N_1, M_1)$, then increasing K_{N_1,M_1} by one (adding the new coefficient $^x\rho_{new}$), which is effective in a new filter $F_a(N_2, M_2)$, follows when and only when*

$$F_a(N_1, M_1) \rightarrow F_a(N_2, M_2) \Leftrightarrow FK(N_1, M_1; N_2, M_2) =$$

$$= \frac{2}{K_{N_1,M_1}} \cdot \frac{^xR_{N_1}^{\{M_1\}}}{|^x\rho_{new}|^2} < \delta \qquad (7)$$

'The partial actualization' by a coefficient $^x\rho_{new}$ can be done by changing N or/and M. This allows to check the different combinations of N and M. The choice for testing the new Schur coefficient depends on the maximal value of $K_{N,M}$. Using proposed strategy 1 the following rules are proposed:

- for determining M increase the parameter N up to reach maximal value $K_{N,M}$; it means the linear, bi-linear, tri-linear,... elements of ON basis are checked, and choosing is one for which $^x\rho_{new}$ meets (1)

Selection the values of N i M is stopped when the desire value of (19) is reached.

4   Simulations

To investigate the reference signal cancellation effect, the described above algorithm was used and compared with the DNLMS algorithm. The two following measures were proposed to establish the quality of the reference signal cancellation process:

– dB_{mean} [dB] - the ratio of the aeroplane echo peak energy on the range doppler surface to the mean value of the energy on this surface except the echo peak,
– dB_{max} [dB] - the ratio of the aeroplane echo peak energy on the range doppler surface to the maximum energy value of the peak which is not the aeroplane echo peak.

Table 1 shows the results of the simulations for the real-life signal.

Table 1. The reference signal cancellation performance - the real-life signal

Filter parameters	DNLMS	NOF^*
N=5, $\mu = 0.5$	dB_{mean}=8.02	dB_{mean}=11.27
	dB_{max}=2.11	dB_{max}=4.64
N=10, $\mu = 0.5$	dB_{mean}=10.40	dB_{mean}=11.64
	dB_{max}=4.55	dB_{max}=4.34
N=20, $\mu = 0.5$	dB_{mean}=11.33	dB_{mean}=12.01
	dB_{max}=5.42	dB_{max}=4.93
N=10, $\mu = 0.05$	dB_{mean}=11.01	dB_{mean}=11.64
	dB_{max}=4.35	dB_{max}=4.34

N - filter order, μ - adaptation coefficient
* order of NOF was selected using Strategy 1 form 1 to N,
M was selected using Strategy 1 from 1 to 3.

First figures show the 'Range-doppler' surface for the radar observed area. Red points mean flying objects.

Next figures present the fragment of the 'Range-doppler' surface near $Range = 150km$ and $Dopplershift = 120Hz$ in 3D style.

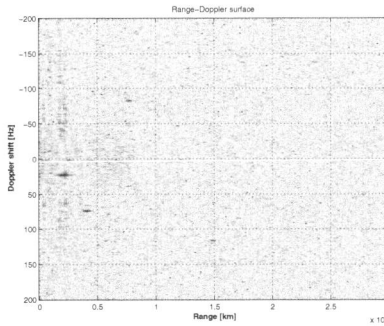

Fig. 4. 'Range-doppler' surface - Lattice filter (N=70)

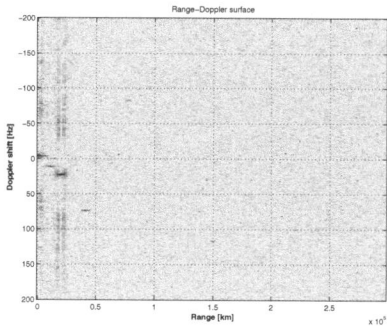

Fig. 5. 'Range-doppler' surface - NOF filter

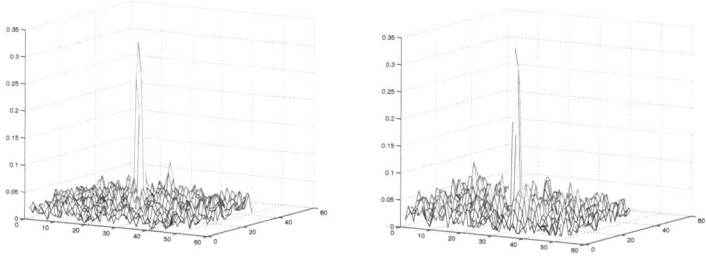

Fig. 6. 'Range-doppler' surface - **Fig. 7.** 'Range-doppler' surface - DNLMS filter (N=70)   NOF

It can be seen how high is the peak of the detected object. The measure of the reference signal cancellation performance were established and is presented in the table 2.

Table 2. The reference signal cancellation performance

DNLMS	NOF
$dB_{mean}=11.10$	$dB_{mean}=11.68$
$dB_{max}=6.02$	$dB_{max}=6.23$

4.1   Eliminating the Echo Signal from Static Objects

To eliminate the echo signal coming from static objects, the presented filters orders must be higher than the maximum range (delay in samples) which is established during the range-doppler surface calculation. For example, for the sampling frequency 195.313e3 Hz and the maximum range 300 km, the filter order must be higher than 195. Such orders require high computing power. This fact was considered during making a decision to choose the reference signal cancellation filter parameters. Choosing the proper coefficients $^x\rho^{j_1,...,j_M}$ (choosing N and M) for NOF and selecting the biggest ones ($^x\rho$), it was possible to control the filter structure to minimize the decorrelation block complexity and get the satisfactory value of the estimation error [2].

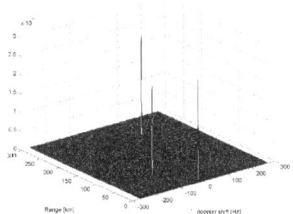

Fig. 8. The range-doppler surface for the generated signal (two objects + one static object): NOF (filter parameters were not sufficient). The filter order smaller than the delay from an object to the radar.

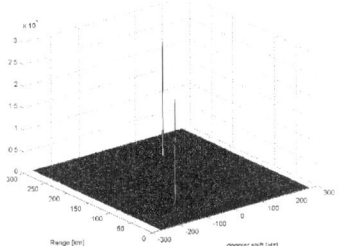

Fig. 9. The range-doppler surface for the generated signal (two objects + one static object): NOF (filter parameters were sufficient). The filter order higher than the delay from an object to the radar.

5 Conclusion

The presented results allow for the following conclusions:

- There are significant differences in the results between the algorithms.
- The change of the filters parameters can improve the cancellation process up to 2 dB.
- The employment of the nonlinear procedures does result in significant improvements due to the fact that the recorded real-life signals are not 'close' to Gaussianity
- The value of the estimation error can be controlled by selecting filter structure during filtering process.

References

1. Biernacki, P., Zarzycki, J.: Multidimensional Nonlinear Noise-Cancelling Filters of the Volterra-Wiener Class. In: Proc. 2-Nd Int. Workshop on Multidimensional (nD) Systems (NDS 2000), pp. 255–261. Inst. of Control and Comp. Eng. TU of Zielona Gora Press, Czocha Castle (2000)
2. Biernacki, P.: Strategies for adaptive nonlinear noise reduction Volterra-Wiener filter structure selection. In: Proc. ICECS 2006, Nice, France, December 10-13 (2006), cd-rom
3. Haykin, S.: Adaptive filter theory. Prentice Hall, Upper Saddle River (1996)
4. Lee, D.T.L., Morf, M., Friedlander, B.: Recursive Least-Squares Ladder Estimation Algorithms. IEEE Trans. on CAS 28, 467–481 (1981)
5. Zarzycki, J.: Multidimensional Nonlinear Schur Parametrization of NonGaussian Stochastic Signals - Part One: Statement of the Problem. Multidimensional Systems and Signal Processing (MDSSP) Journ. 15(3), 217–241 (2004)
6. Biernacki, P.: A new strategy of adaptive nonlinear echo cancelling Volterra-Wiener filter structure selection. In: Setchi, R., Jordanov, I., Howlett, R.J., Jain, L.C. (eds.) KES 2010, Part I. LNCS, vol. 6276, pp. 350–359. Springer, Heidelberg (2010)

A Web Browsing Cognitive Model

Pablo E. Román* and Juan D. Velásquez

Department of Industrial Engineering, República 701, Santiago, Chile
proman@ing.uchile.cl, jvelasqu@dii.uchile.cl

Abstract. Web usage have been studied from the point of view of machine learning. Although web usage prediction are mostly restricted to an static web site structure, hence this results to be a hard restriction to accomplish in the practice. We propose a decision-making model that allow predicting web users' navigation choices even in dynamics web sites. We propose a neurophysiological theory of web browsing decision making, which is based on the Leaky Competing Accumulator (LCA). The model is stochastic and has been studied in the context of Psychology for many years. Choices are performed to follow hyperlink according to user text preferences. This process is repeated until the web user decide to leave the web site. Model's parameters are required to be fitted in order to perform Monte Carlo simulations. It has been observed that nearly 73% of the real distribution is recovered by this method.

Keywords: Neurocomputing, Web User Behavior, LCA, Neurophysiology, Stochastic Equation, Stochastic Simulation, Text Preferences, Web Session, Curse of Dimensionality, Markov processes.

1 Introduction

The present work describes a novel approach by applying a neurophysiological theory of decision making for describing web user browsing behavior. The research hypothesis is: It is possible to apply neurophysiologys decision making theories to explain web user navigational behavior using web data. The current approach focuses on a narrow class of individual web users. Several models from Mathematical psychology have been presented on this subject, but few have so far been applied to the engineering field. Far from being a complete description of the conscious process of decision making, this theory describes how much time the subject will take to make a determination based on preconceived likelihood of each option. An objective of this research is to predict changes in the navigational behavior of the web user based on historical data. Furthermore, the Web User/Site system is a human-machine interaction system.

While current approaches for studying the web user's browsing behavior are based on generic machine learning approaches, a rather different point of view is developed and presented here. A model based on the neurophysiology theory of decision making is applied to the link selection process. This model has two stages,

---

* Corresponding author.

M. Graña et al. (Eds.): KES 2012, LNAI 7828, pp. 31–40, 2013.
© Springer-Verlag Berlin Heidelberg 2013

the training stage and the simulation stage. In the first, the model's parameters are adjusted to the user's data. In the second, the configured agents are simulated within a web structure for recovering the expected behavior. The main difference with the machine learning approach consists in that the model is independent of the structure and content of the web site. Furthermore, agents can be confronted with any page and decide which link to follow (or leave the web site). This important characteristic makes this model appropriate for heavily dynamic web sites. Therefore, such an agent model may operates on any changing environment. Another important difference is that the model has a strong theoretical basis built upon physical phenomenon. Traditional approaches are generic, but this proposal is based on a state-of-the-art theory of brain decision making.

The rest of the paper is organized as follows. Section 2 offers a description of the neuro-physiological model for decision making. Section 3 is related to the application of the model to the surfing process. In section 4 the mathematics and algorithm of the simulation are described. In section 5 the parameter adjustment is revised. Section 6 shows the numerical results of experiments. Finally, section 7 states the main conclusions of this study.

2   The Leaky Competing Accumulator Model (LCA)

The proposal is based on the model named LCA (Leaky Competing Accumulator) [1]. This model associates the neural activity levels of certain brain regions in the lateral intra parietal cortex (LIP) with a discrete set of possible choices. Such values relate to average electrical signal rate and are measured in spikes per second. Those (X_i) evolve according to a stochastic equation (1) during the agent's decision-making process until one of the activity values reaches a given threshold. Such an equation describes the small changes dX_i that affect a neural activity X_i associated with an option i, after a small time step dt. Individuals act accumulating evidence $(X = \int dX)$, summing up said increments for each possible choice until one of the neuronal activity levels i is considered high enough to effect a decision. The first coordinate to hit the barrier $\Psi = \{X | \exists i, \ X_i = 1\}$ indicates which associated decision is made.

$$dX_i = (-\kappa X_i - \lambda \sum_{j \neq i} X_j + \beta I_i)dt + \sigma dW_i \qquad (1)$$

Models like the LCA stochastic process have a long history of research and experimental validations, most of which have been carried out in the last 40 years (see [2–4]). This particular model intends to describe real neural activities, in contrast to others approaches that use variables unrelated to physical phenomena. However, few engineering applications have been proposed until now. This paper assumes that those proven theories regarding human behavior can be applied and adapted to describe web user behavior, producing a more effectively structured and specific machine learning model. The stochastic process (equation 1) depends strongly on the choice evidence levels parameter vector I. A value (I_i) (equation 1) is a neural activity level of a brain region that is associated with a unique choice, and whose value anticipates the likelihood for the choice before the decision is made. A 2002 experiment on rhesus monkeys revealed how decisions based on visual stimuli correlate with the middle temporal area (MT) of the brain [5] that corresponds to I_i parameters. The MT area appears to be a temporary storage and processing device for visual information that is transmitted to the LIP area for further processing. Earlier studies [6] showed a linear correlation of visual stimuli on the neural activity of the visual cortex MT. For decision making, the brain seems to integrate information processed from other areas of the brain $\{I_j\}$ and triggers a determination (i) when the neural activity (X_i) reaches a certain threshold. The LCA model needs a specification of the $\{I_i\}$ parameters in order to simulate a web user browsing decision, as described in the next section.

3   Web Browsing Decision-Making Model

Web users are modeled as information foragers [7]. Furthermore, they experience a degree of satisfaction with consuming the information included on web pages. This idea is influenced by the economic theory of utility maximization, where a web user is a consumer of information, selecting the link that most satisfies him/her. However, a model which only considers this dynamic factor would produce a web user that never stops navigating. The Random Surfer [8] Model is a naive description of a web user that has no interest at all in the web page content. Furthermore, a random surfer does not rely on web page content, he/she only uniformly decides on the next link to follow, or leaves the site with probability d. The proposed model combined both paradigm using the LCA decision making framework. Web users are considered stochastic agents [9–12]. Those agents follow LCA stochastic model dynamics (Equation 1), and maintain an internal state X_i affected by white noise dW_i and other interactions in equation 1. The available choices, including the probability of leaving the web site, lie in the links on a web page. Agents make decisions according to their internal preferences using a utilitarian scheme. The values (I) are the main forces that drive the decision system (1). Furthermore, we model those values as proportional to the probability $P(i)$ of the discrete choices $(I_i = \beta P(i))$, which are usually modeled using the Random Utility Model. Discrete choice preferences have been studied in economics to describe the amount of demand for discrete goods where consumers are considered rational as utility maximizers. The utility maximization problem regarding discrete random variables results in a class of extreme probability distributions, in which the widely-used model is the random utility model $(P(i) = e^{V_i} / \sum_j e^{V_j})$ and where probabilities are adjusted using the known logistics regression [13]. The probability distribution of a choice i anticipates every possible choice on the page j and has a consumer utility $V_j + \epsilon_i$, where ϵ_i is the random part of the utility. This technique has successfully been applied to modeling a user's search for information on a hypertext system [7], resulting in improved adaptive systems. The utility function should depend on the text present in links that the user then interprets and by means of which he/she makes the decision. Nevertheless, this extension of the LCA model is much more general since it can be applied to other fields by using the correct discrete choice utility. In this case the utility depends on perceived information, assuming it is mainly based on text. The assumption is that each agent's link preferences are defined by its TF-IDF text vector $\mu = [\mu_k]$ [14]. The TF-IDF weight μ_k component is interpreted as the importance to the web user of the word k. Furthermore, an agent prefers to follow similar links to its vector μ. The utility values (equation 2) are given by the dot product between the normalized TF-IDF vector μ and L_i that represents the TF-IDF weight text vector associated with the link i.

$$V_i(\mu) = \mu \bullet L_i / |\mu||L_i| \tag{2}$$

The resulting stochastic model (equation 1) is dependent on the parameters $\{\kappa, \lambda, \sigma, \beta, \mu\}$. The set of vectors $\{L_i\}$ are exogenous parameters that depend on the particular web site structure and content. The first four parameters could be considered as constants for all users, yet the μ vector should an intrinsic characteristic of each. In this sense, the real web user's mode of behavior as observed on a web site corresponds to a distribution of users. We are assuming the web user does not change his/her intention during a session and leave the web site according to constant probability. Therefore, in this model web user profiling corresponds with the μ vector probability distribution.

4   Simulation

The Web User/Web Site system is rather complex since it incorporates human behavior into often dynamically-generated pages. At first sight a complete mathematical description of the individual session evolution would seem to be far to accomplish. We use "Stochastic Simulation" since some of the variables of the system are random for recovering the web user navigational behavior. The following algorithm describe a simplified version of the process.

Simulation of web usage is implemented using two main levels. The first level is driven by navigational decisions on pages, recovering a simulated visit trail (Algorithm 1). The second level consists of finding the choice of the next navigational action and the time taken for it (Algorithm 2). Nevertheless, the decision of which navigational operation the user will take is driven by the LCA model. This process is repeated for obtaining an approximation of statistical values. Nowadays, Monte Carlo techniques are simpler and easier to implement, yet this method has slow convergence rates. Nevertheless, this procedure should be executed on several threads in order to ensure faster statistical convergence. This single-agent procedure should be executed with different text vector preferences for recovering the associated different modes of behavior on the web site. The navigational dynamic of choice is proposed to be described by the LCA stochastic process. A schema of processing the hyperlink choice is detailed in the following algorithm. The one-step iteration is improved using exact simulation [15]. Furthermore, the LCA process without considering border condition can be integrated exactly and it is called the Ornstein-Uhlenbeck process ([12, 16]). Such an exact path solution from the stochastic equation [12] should be discretized for reproducing procedure 2 until a boundary condition is reached.

Algorithm 1. Web navigation simulation: The session's generation

---

1: **Select a random initial page** p according to the observed empirical distribution of the first session pages. Probabilities are considered by the observed
   rate of hit per initial page.
2: **Initialize** the session sequence $S \leftarrow \{\}$.
3: **repeat**
4:     **Simulate the web user's next hyperlink selection**, obtaining the
       next page p' and the time τ used for the decision (Algorithm 2).
5:     **Push** the current page on the session set $S \leftarrow S \cup \{(p, \tau)\}$
6:     **Select** the new page $p \leftarrow p'$
7: **until** the page p **NOT** corresponds to the sink node.
8: **Return** $S \cup \{(p, \tau)\}$

---

5   Parameter Estimation

On the other hand, simulation requires all of its parameters to have definite values in order to recover predictability. The process of finding such values is called fitting, and the objective is to adjust the model to recover as much of the observed behavior as possible. In this case, observed web user's sessions are used for reaching a maximal likelihood. Moreover, the web user text preference vector μ is considered to be described by a multivariate distribution. Web users that visit a web site are described by a variety of objectives defined by the μ distribution. Finding the distribution of vector I is a difficult task. First of all, such a distribution does not relate with text choice since it reflects the distribution of different kinds of subjects classified by their text preference. However, two assumptions involve further simplification of the inference process. Similar sessions should group similar web user text preferences and distribute like-multivariate normal distributions within each cluster ζ. This method has been explored [17] for web user simulation by ant colony optimization. Our methods differs from previous work [11, 12] in fitting the model by cluster in similar way to [17]. Cluster set $\{\zeta\}$ must be obtained by hierarchical clustering techniques, given the discrete character of the similarity between trails. The reason is that there is no way to define a middle point between two sessions. Similarity relates with the size $|\mathrm{MCS}(s_1, s_2)|$ of the maximal common subsequence [18] in the sense that sessions s_1, s_2 with a maximal degree of common path are considered. Finally the similarity function between s_1 and s_2 is defined by $2|\mathrm{MCS}(s_1, s_2)|/(|s_1| + |s_2|)$. Transitions are then segregated by the identified clusters, expectancy and variance for the restricted set $\bar{I}_\zeta, \Sigma_\zeta$ are found by the optimization procedure. The difficulty of having to estimate a distribution as a parameter was described. Using a kind of average $\bar{I}$

Algorithm 2. Simple Simulation of a Navigational Decision in a page p

---
1: **Initialize** the vector $X = 0$, which have the same dimension as different links in the page p. It includes a component representing the option for leaving the website (sink node).
2: **Initialize** the user text preference vector μ and prepare the vector L_i as the TF-IDF text vector associated with each link i on page p.
3: **Evaluate** the vector $I = [I_i]$ using the similarity measure with μ and the text vector L_i associated with the link i as shown in equation 2. In this case we approximate L_i by the TF-IDF values of the pointed-to page.
4: $\tau \leftarrow 0$
5: **repeat**
6:    **Perform one-step iteration** for the evolution of the vector X. The Euler method uses a time step h as shown in equation 1 and is the most-used technique for this step.
7:    **if** one or more $\{X_i\}$ components become negative **then**
8:       **reset them to** 0.
9:    **end if**
10:   $\tau \leftarrow \tau + h$
11: **until** The threshold is reached.
12: $j' \leftarrow Argmax_i\{X_i\}$
13: **Return** the simulated choice is j' and the time taken for the decision τ.

---

as a partial solution could be a way for finding the calibration of the model. Maximum likelihood is a well-known technique for stochastic model calibration. The probability of observing the available data (likelihood) is maximized using as variables the unknown parameters, subject to the restrictions of the theoretical model. The solution is interpreted to be optimal in the sense of being the most probable according to the observation. The calculation for obtaining a kind of average $\bar{I}$ vector is based on observed data from a real web site. For a given possible transition from the page i to j, a number n_{ijk} of observed clicks measures the time spent on the website t_{ijk}. Despite the presence of an exact polynomial solution [12] for the unconstrained probability function, it is difficult to obtain an explicit expression for the likelihood expression. It turns out much more unmanageable to calculate probabilities by partitioning the domain of the problem. Indeed if each dimension (possible choices) is partitioned in 100 parts and there are 20 links on a page (typical number), then the number of points turns out to be an astronomical 100^{20}. This problem is called "the curse of dimensionality" since computational complexity explodes with dimensionality. This issue is common for many differential problems related to the neural decision framework. Fortunately, a much more computationally inexpensive algorithm is proposed for calculating such a function. Symbolic integration can manage very efficiently the computation of the likelihood. A first observation is that the multivariate integrated function is a polynomial on variables $\{X, I\}$ and a rational function on $\{\kappa, \lambda, \beta, \sigma\}$, if likelihood is approximated by polynomials. As such, integrals over variable X are straightforwardly calculated and evaluated exactly by symbolic integration. This observation is important since it drastically reduces the computational complexity of the inference algorithm. Furthermore, derivatives of the likelihood function on I, σ, κ, λ, and β can be directly extracted after symbolic processing. In this way, traditional non-linear optimization methods can be used in this system using the resulting evaluation of the maximized function. Nevertheless, the time-dependent probability function needs to satisfy the

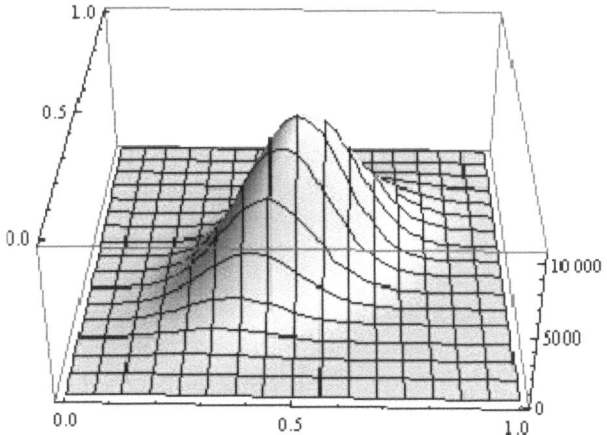

Fig. 1. Polynomial approximation shape (degree= 100) of the probability distribution for two choices and $I = (1, 1)$ in a given time t. Parameters are $\kappa = 0.8, \lambda = 0.3, \sigma = 0.2$.

differential problem 1, border conditions, and be described by polynomials or rational functions. Since exact solutions are not available, approximation can be performed using a polynomial (or rational) initial solution accumulated near 0 and using the Fokker-Plank propagator for finding the solution at any time t [12]. The shape of the resulting probability function is illustrated in figure 1. In this way, any calculation to obtain the likelihood involves partial derivatives and integration operations that can be straightforwardly calculated by symbolic processing. The resulting likelihood function is now calculated in polynomial time avoiding the dimensionality-explosion problem.

6   Results

In this research we used five sub-sites belonging to the Industrial Engineering Department (http://www.dii.uchile.cl). The main departmental site, three sub-sites from the master's degree program, and a project web site have nearly a thousand web pages. Each one has its own characteristics in terms of content and structure and there is no homogeneity in relation to the process of web construction. This web site has the property of having a lower degree of complexity, and changes in the web site are minimal compared with others. Those characteristics make sessions on average simpler and ideal for the study of web usage mining. Real sessions are retrieved by mean of a cookie method and compared with simulations. The data retrieval process started in June of 2009 and finished in August 2010. Web user sessions were obtained through a cookie-based sessionization method, which stores the anonymous user identifier. This technique uses both client-side and server-side elements to track web user behavior and automatically reconstruct the sessions. The following result was obtained; $\lambda : 0.4$, $\kappa : 0.2$, $\sigma : 0.03$, $\beta : 0.45$. Those parameters were then fixed for performing the calibration of the evidence vector. The symbolic-based algorithm was stated and a single vector was fit. An interesting fact is that the result was nearly similar to the vector obtained by the most important word in the web site. An error of 40% is identified in the session distribution simulation. The process takes nearly 10 hours. A distribution of vector was obtained clustering the session by using the longest common subsequence distance for clustering. In this the clustering process was stopped when 10 clusters were detected. The same previous process of calibration was performed using this method on each cluster and subset of visited pages. The results were astonishing, nearly $8, 3\%$ of error were obtained in the simulated session distribution. As we have seen session length follows a typical distribution [19]. The simulated "average web user" follows a distribution of session length similar to the empirical. The relative error is of $8, 1\%$, or less than 1% of error in log scale. Distribution error remains more or less constant for sessions with a capacity of less than 15 pages consisting of $0, 3\%$ of error. The regression results obtained from the distribution of the μ text preference vector show that most probably (maximal $P(u)$) have the following word highly ranked. Three vectors were selected on the upper 70% of probability. A sample set of the words obtained by this method is presented:

1. **Management, Engineering, Enterprise, Society**. Interpretation: related to the industrial engineering field.
2. **Mgpp, Council, Bank, Description. Interpretation**: Word related to a master in public politics degree (Mgpp).
3. **Capital, Market, Sustainability, Characterization**. Interpretation: Word related to economics.

Those results show great accuracy in describing the interest of a real visitor to the web site of the Industrial Engineering Department of the University of Chile. It was known that traditional machine learning results in $50 - 70\%$ of effectiveness for rebuilding the distribution of session [20]. However, such algorithms only predict the next step of the session based on previous history. Indeed, if considering machine learning predict next step of a session by nearly 70% of the cases, then the whole sessions would be recovered geometrically in a fraction much less than this value. The current model differs from the capability of generating entire sessions samples. The effectiveness of this method is nearly 78.5% of the generated session distributed like the real one. This is calculated from the frequency of appearance of each real session corresponding to the average of the recovered fraction of them. The most important testing of the model is related to future behavior based on past training. The effectiveness of this method in the next two months after calibration was of about 73%.

7   Conclusion

This work has proposed to study and apply a psychology-based theory of decision making to web usage in order to describe a user's sessions. The mechanism used for such purposes was the neurophysiology LCA (equation 1) model of decision making. It is adapted to predict the next page in a session and obtaining the sequence of visited pages. This corresponds to a stochastic simulation scheme that is handled by Monte Carlo techniques. Once the model is calibrated, it is possible to obtain the distribution navigation trails using Monte Carlo techniques. As a sub-product of the calibration mechanism it results in the dispersion of the web user's keyword interest. A major disadvantage of the model is the mathematical difficulty of approximating solutions. The problem originates in the naturally high number of dimensions for the differential problem (typically 20) and non-standard border conditions. Differential problems are commonly solved by a discrete mesh on the domain are computationally intractable in this case. In spite of the intricate mathematical description, the model is based on physiological principles of decision making validated experimentally. This characteristic suffices for developing and exploring the capabilities of this model by means of adjusting the theoretical dynamics to the observed fact. An approach based on symbolic processing and an exact polynomial solution of the unconstrained problem was proposed to avoid a dimensional explosion. Our presented model differ from [11, 12] on the simplifying assumption on fitting the model on clustered session [17]. Since the simulation successfully predicts navigation changes, then changes on a web site can be investigated for optimizing measures of web site usability. Such a possibility drastically alters the current concept of an adaptive web site since it is possible to predict the impact of a change on the web site. Web page semantic has not been considered in the model, since the bag of word model dismisses those relations. Visual disposition and presentation of hyperlink on a web page, could also influence the decision about the next page to visit. This phenomena should be included in the I vector, since in this case, relates with brains visual processing and semantic resolution. In the proposed framework a utility function resume preferences. Text semantic and/or visual disposition, should be represented by a numeric vector, so it enters on the utility definition.

Acknowledgment. This work was partially supported by the FONDEF project D10I1198 and the Millennium Institute on Complex Engineering Systems (ICM: P-05-004-F, CONICYT: FBO16).

References

1. Usher, M., McClelland, J.: The time course of perceptual choice: The leaky, competing accumulator model. Psychological Review 2(1), 550–592 (2001)
2. Laming, D.R.J.: Information theory of choice reaction time. Wiley (1968)
3. Ratcliff, R.: A theory of memory retrieval. Psychological Review 85(2), 59–108 (1978)
4. Busemeyer, J.R., Jessup, R.K., Johnson, J.G., Townsend, J.T.: Building bridges between neural models and complex decision making behaviour. Neural Networks 19(8), 1047–1058 (2006)
5. Roitman, J.D., Shadlen, M.N.: Response of neurons in the lateral intraparietal area during a combined visual discrimination reaction time task. Journal of Neuroscience 22, 9475–9489 (2002)
6. Britten, K.H., Shadlen, M.N., Newsome, W.T., Movshon, J.A.: Response of neurons in macaque motion signals. Visual Neuroscience 10, 1157–1169 (2006)
7. Pirolli, P.: Power of 10: Modeling complex information-seeking systems at multiple scales. Computer 42, 33–40 (2009)
8. Blum, A., Chan, T.H.H., Rwebangira, M.R.: A random-surfer web-graph model. In: Procs. ALENEX8 SIAM, pp. 238–246. SIAM (2006)
9. Román, P.E., Velásquez, J.D.: A dynamic stochastic model applied to the analysis of the web user behavior. In: The 2009 AWIC 6th Atlantic Web Intelligence Conference, pp. 31–40 (2009)
10. Román, P.E., Velásquez, J.D.: Analysis of the web user behavior with a psychologically-based diffusion model. In: The AAAI 2009 Fall Symp. BICA (2009)
11. Román, P.E., Velásquez, J.D.: Stochastic simulation of web users. In: Procs. of the 2010 WIC (September 2010)
12. Román, P.E.: Web User Behavior Analysis. PhD thesis, U. of Chile (2011)
13. Newey, W., McFadden, D.: Large sample estimation and hypothesis testing. Handbook of Econometrics 4, 2111–2245 (1994)
14. Manning, C.D., Schutze, H.: Fundation of Statistical Natural Language Processing. The MIT Press (1999)
15. Gillespie, D.T.: Exact numerical simulation of the ornstein-uhlenbeck process and its integral. Physical Review E 54(2), 2089–2091 (1996)
16. Beskos, A., Papaspiliopoulos, O., Roberts, G.O.: Retrospective exact paths with applications. Bernoulli 12(6), 1077–1098 (2006)
17. Loyola, P., Román, P.E., Velásquez, J.D.: Predicting web user behavior using learning-based ant colony optimization. Eng. Appl. of AI 25(5), 889–897 (2012)
18. Spiliopoulou, M., Mobasher, B., Berendt, B., Nakagawa, M.: A framework for the evaluation of session reconstruction heuristics in web-usage analysis. Informs JOC 15(2), 171–190 (2003)
19. Huberman, B., Pirolli, P., Pitkow, J., Lukose, R.M.: Strong regularities in world wide web surfing. Science 280(5360), 95–97 (1998)
20. Velasquez, J., Palade, V.: A knowledge base for the maintenance of knowledge. Journal of Knowledge Based Systems 1(20), 238–248 (2007)

Optimization of Approximate Decision Rules Relative to Number of Misclassifications: Comparison of Greedy and Dynamic Programming Approaches

Talha Amin[1], Igor Chikalov[1], Mikhail Moshkov[1], and Beata Zielosko[1,2]

[1] Computer, Electrical and Mathematical Sciences and Engineering Division
King Abdullah University of Science and Technology
Thuwal 23955-6900, Saudi Arabia
{talha.amin,igor.chikalov,mikhail.moshkov}@kaust.edu.sa,
beata.zielosko@us.edu.pl
[2] Institute of Computer Science, University of Silesia
39, Będzińska St., 41-200 Sosnowiec, Poland

Abstract. In the paper, we present a comparison of dynamic programming and greedy approaches for construction and optimization of approximate decision rules relative to the number of misclassifications. We use an uncertainty measure that is a difference between the number of rows in a decision table T and the number of rows with the most common decision for T. For a nonnegative real number γ, we consider γ-decision rules that localize rows in subtables of T with uncertainty at most γ. Experimental results with decision tables from the UCI Machine Learning Repository are also presented.

Keywords: decision rules, number of misclassifications, dynamic programming algorithm, greedy algorithm.

1 Introduction

Decision rules are used in many areas connected with data mining and machine learning. The number of misclassifications is important parameter if we consider decision rules as a way for construction of classifier [8–10, 12, 13].

There are many approaches to the construction of decision rules, for example, Apriori algorithm [1], different kinds of greedy algorithms [9, 11], dynamic programming [3–6, 14]. This paper, extending a conference publication [4], is devoted to the comparison of the number of misclassifications of decision rules constructed by dynamic programming algorithm and greedy algorithm. We use as an uncertainty measure $J(T)$ that is a difference between the number of rows in a given decision table and the number of rows labeled with the most common decision in this table. We fix a nonnegative threshold γ, and study γ-decision rules that localize rows in subtables whose uncertainty is at most γ. For each of such rules the number of misclassifications is at most γ. In [4] we presented

M. Graña et al. (Eds.): KES 2012, LNAI 7828, pp. 41–50, 2013.

a dynamic programming algorithm for decision rule optimization relative to the number of misclassifications. In [3] we studied dynamic programming approach for partial decision rule optimization relative to the length and coverage. In [14] we discussed possibilities of sequential optimization of γ-decision rules relative to the length, coverage and number of misclassifications.

Dynamic programming approach allows us to find optimal (from different points of view) decision rules. In this paper, we present dynamic programming algorithm that allows us to obtain γ-decision rules with the minimum number of misclassifications. The presented algorithm constructs a directed acyclic graph $\Delta_\gamma(T)$. Based on this graph we can describe the whole set of so-called irredundant γ-decision rules. Then optimize rules from this set relative to the number of misclassifications. We also present a greedy algorithm for construction of γ-decision rules. It allows us to make some comparative study relative to the number of misclassifications for constructed approximate decision rules.

This paper consists of seven sections. Section 2 contains definitions of main notions. In Sect. 3, we study a directed acyclic graph which allows us to describe the whole set of irredundant γ-decision rules. In Sect. 4, we consider a procedure of optimization of irredundant γ-decision rules relative to the number of misclassifications. In Sect. 5, we present a greedy algorithm for construction of γ-decision rules. Section 6 contains results of experiments with decision tables from the UCI Machine Learning Repository [7]. Section 7 contains conclusions.

2   Main Notions

In this section, we consider definitions of notions corresponding to decision tables and decision rules.

A *decision table* T is a rectangular table with n columns labeled with conditional attributes $f_1, \ldots, f_n$. Rows of this table are filled by nonnegative integers which are interpreted as values of conditional attributes. Rows of T are pairwise different and each row is labeled with a nonnegative integer (decision) which is interpreted as value of the decision attribute. It is possible that T is empty, i.e., has no rows.

A minimum decision value which is attached to the maximum number of rows in T will be called the *most common decision for* T. The most common decision for empty table is equal to 0.

We denote by $N(T)$ the number of rows in the table T and by $N_{mcd}(T)$ we denote the number of rows in the table T labeled with the most common decision for T. We will interpret the value $J(T) = N(T) - N_{mcd}(T)$ as *uncertainty* of the table T.

The table T is called *degenerate* if T is empty or all rows of T are labeled with the same decision. It is clear that $J(T) = 0$ if and only if T is a degenerate table.

A table obtained from T by the removal of some rows is called a *subtable* of the table T. Let T be nonempty, $f_{i_1}, \ldots, f_{i_k} \in \{f_1, \ldots, f_n\}$ and $a_1, \ldots, a_k$ be nonnegative integers. We denote by $T(f_{i_1}, a_1) \ldots (f_{i_k}, a_k)$ the subtable of the table T which contains only rows that have numbers $a_1, \ldots, a_k$ at the intersection with columns $f_{i_1}, \ldots, f_{i_k}$. Such nonempty subtables (including the table T) are called *separable subtables* of T.

We denote by $E(T)$ the set of attributes from $\{f_1, \ldots, f_n\}$ which are not constant on T. For any $f_i \in E(T)$, we denote by $E(T, f_i)$ the set of values of the attribute f_i in T.

The expression

$$f_{i_1} = a_1 \wedge \ldots \wedge f_{i_k} = a_k \to d \qquad (1)$$

is called a *decision rule over* T if $f_{i_1}, \ldots, f_{i_k} \in \{f_1, \ldots, f_n\}$, and $a_1, \ldots a_k, d$ are nonnegative integers. It is possible that $k = 0$. In this case (1) is equal to the rule

$$\to d. \qquad (2)$$

Let $r = (b_1, \ldots, b_n)$ be a row of T. We will say that the rule (1) is *realizable for* r, if $a_1 = b_{i_1}, \ldots, a_k = b_{i_k}$. If $k = 0$ then the rule (2) is realizable for any row from T.

Let γ be a nonnegative real number. We will say that the rule (1) is γ-*true for* T if d is the most common decision for $T' = T(f_{i_1}, a_1) \ldots (f_{i_k}, a_k)$ and $J(T') \leq \gamma$. If $k = 0$ then the rule (2) is γ-true for T if d is the most common decision for T and $J(T) \leq \gamma$.

If the rule (1) is γ-true for T and realizable for r, we will say that (1) is a γ-*decision rule for* T *and* r.

We will say that the rule (1) with $k > 0$ is an *irredundant* γ-decision rule for T and r if (1) is a γ-decision rule for T and r and the following conditions hold:

(i) $f_{i_1} \in E(T)$, and if $k > 1$ then $f_{i_j} \in E(T(f_{i_1}, a_1) \ldots (f_{i_{j-1}}, a_{j-1}))$ for $j = 2, \ldots, k$;
(ii) $J(T) > \gamma$, and if $k > 1$ then $J(T(f_{i_1}, a_1) \ldots (f_{i_j}, a_j)) > \gamma$ for $j = 1, \ldots, k-1$.

If $k = 0$ then the rule (2) is an *irredundant* γ-decision rule for T and r if (2) is a γ-decision rule for T and r, i.e., if d is the most common decision for T and $J(T) \leq \gamma$.

Let τ be a decision rule over T and τ be equal to (1). The *number of misclassifications* of τ is the number of rows in T for which τ is realizable and which are labeled with decisions different from d. We denote it by $\mu(\tau)$. The number of misclassifications of the decision rule (2) is equal to the number of rows in T which are labeled with decisions different from d.

3   Directed Acyclic Graph $\Delta_\gamma(T)$

Now, we consider an algorithm that constructs a directed acyclic graph $\Delta_\gamma(T)$ which will be used to describe the set of irredundant γ-decision rules for T and for each row r of T. Nodes of the graph are separable subtables of the table T. During each step, the algorithm processes one node and marks it with the symbol *. At the first step, the algorithm constructs a graph containing a single node T which is not marked with the symbol *.

Let the algorithm have already performed p steps. Let us describe the step $(p+1)$. If all nodes are marked with the symbol * as processed, the algorithm finishes its work and presents the resulting graph as $\Delta_\gamma(T)$. Otherwise, choose a node (table) Θ, which has not been processed yet. Let d be the most common decision for Θ. If $J(\Theta) \leq \gamma$ label the considered node with the decision d, mark it with symbol * and proceed to the step $(p+2)$. If $J(\Theta) > \gamma$, for each $f_i \in E(\Theta)$, draw a bundle of edges from the node Θ. Let $E(\Theta, f_i) = \{b_1, \ldots, b_t\}$. Then draw t edges from Θ and label these edges with pairs $(f_i, b_1), \ldots, (f_i, b_t)$ respectively. These edges enter to nodes $\Theta(f_i, b_1), \ldots, \Theta(f_i, b_t)$. If some of nodes $\Theta(f_i, b_1), \ldots, \Theta(f_i, b_t)$ are absent in the graph then add these nodes to the graph. We label each row r of Θ with the set of attributes $E_{\Delta_\gamma(T)}(\Theta, r) = E(\Theta)$. Mark the node Θ with the symbol * and proceed to the step $(p+2)$. The graph $\Delta_\gamma(T)$ is a directed acyclic graph. A node of such graph will be called *terminal* if there are no edges leaving this node. Note that a node Θ of $\Delta_\gamma(T)$ is terminal if and only if $J(\Theta) \leq \gamma$.

Later, we will describe the procedure of optimization of the graph $\Delta_\gamma(T)$ relative to μ. As a result we will obtain a graph $\Delta_\gamma(T)^\mu$ with the same sets of nodes and edges as in $\Delta_\gamma(T)$. The only difference is that any row r of each nonterminal node Θ of $\Delta_\gamma(T)^\mu$ is labeled with a nonempty set of attributes $E_{\Delta_\gamma(T)^\mu}(\Theta, r) \subseteq E(\Theta)$.

Let G be either the graph $\Delta_\gamma(T)$ or the graph $\Delta_\gamma(T)^\mu$. Now, for each node Θ of G and for each row r of Θ, we describe the set of γ-decision rules $Rul_G(\Theta, r)$. We will move from terminal nodes of G to the node T.

Let Θ be a terminal node of G labeled with the most common decision d for Θ. Then $Rul_G(\Theta, r) = \{\rightarrow d\}$.

Let now Θ be a nonterminal node of G such that for each child Θ' of Θ and for each row r' of Θ', the set of rules $Rul_G(\Theta', r')$ is already defined. Let $r = (b_1, \ldots, b_n)$ be a row of Θ. For any $f_i \in E_G(\Theta, r)$, we define the set of rules $Rul_G(\Theta, r, f_i)$ as follows:

$$Rul_G(\Theta, r, f_i) = \{f_i = b_i \wedge \sigma \rightarrow s : \sigma \rightarrow s \in Rul_G(\Theta(f_i, b_i), r)\}.$$

Then $Rul_G(\Theta, r) = \bigcup_{f_i \in E_G(\Theta, r)} Rul_G(\Theta, r, f_i)$.

Theorem 1. *[3] For any node Θ of $\Delta_\gamma(T)$ and for any row r of Θ, the set $Rul_{\Delta_\gamma(T)}(\Theta, r)$ is equal to the set of all irredundant γ-decision rules for Θ and r.*

An example of the presented algorithm work can be found in [4].

4   Procedure of Optimization Relative to Number of Misclassifications

Let $G = \Delta_\gamma(T)$. We consider the procedure of optimization of the graph G relative to the number of misclassifications μ. For each node Θ in the graph G, this procedure corresponds to each row r of Θ the set $Rul_G^\mu(\Theta, r)$ of γ-decision rules with minimum number of misclassifications from $Rul_G(\Theta, r)$ and the number $Opt_G^\mu(\Theta, r)$ – the minimum number of misclassifications of a γ-decision rule from $Rul_G(\Theta, r)$.

The idea of the procedure is simple. It is clear that for each terminal node Θ of G and for each row r of Θ, the following equalities hold:

$$Rul_G^\mu(\Theta, r) = Rul_G(\Theta, r) = \{\to d\},$$

where d is the most common decision for Θ, and $Opt_G^\mu(\Theta, r)$ is equal to the number of rows in Θ labeled with decisions different from d.

Let Θ be a nonterminal node of G, and $r = (b_1, \ldots, b_n)$ be a row of Θ. We know that

$$Rul_G(\Theta, r) = \bigcup_{f_i \in E_G(\Theta, r)} Rul_G(\Theta, r, f_i)$$

and, for $f_i \in E_G(\Theta, r)$,

$$Rul_G(\Theta, r, f_i) = \{f_i = b_i \wedge \sigma \to s : \sigma \to s \in Rul_G(\Theta(f_i, b_i), r)\}.$$

For $f_i \in E_G(\Theta, r)$, we denote by $Rul_G^\mu(\Theta, r, f_i)$ the set of all γ-decision rules with the minimum number of misclassifications from $Rul_G(\Theta, r, f_i)$ and by $Opt_G^\mu(\Theta, r, f_i)$ we denote the minimum number of misclassifications of a γ-decision rule from $Rul_G(\Theta, r, f_i)$.

One can show that

$$Rul_G^\mu(\Theta, r, f_i) = \{f_i = b_i \wedge \sigma \to s : \sigma \to s \in Rul_G^\mu(\Theta(f_i, b_i), r)\},$$

$$Opt_G^\mu(\Theta, r, f_i) = Opt_G^\mu(\Theta(f_i, b_i), r),$$

and $Opt_G^\mu(\Theta, r) = \min\{Opt_G^\mu(\Theta, r, f_i) : f_i \in E_G(\Theta, r)\} = \min\{Opt_G^\mu(\Theta(f_i, b_i), r) : f_i \in E_G(\Theta, r)\}$. It's easy to see also that

$$Rul_G^\mu(\Theta, r) = \bigcup_{f_i \in E_G(\Theta, r), Opt_G^\mu(\Theta(f_i, b_i), r) = Opt_G^\mu(\Theta, r)} Rul_G^\mu(\Theta, r, f_i).$$

We now describe the procedure of optimization of the graph G relative to the number of misclassifications μ.

We will move from the terminal nodes of the graph G to the node T. We will correspond to each row r of each table Θ the number $Opt_G^\mu(\Theta, r)$ which is the minimum number of misclassifications of a γ-decision rule from $Rul_G(\Theta, r)$ and we will change the set $E_G(\Theta, r)$ attached to the row r in Θ if Θ is a nonterminal node of G. We denote the obtained graph by G^μ.

Let Θ be a terminal node of G and d be the most common decision for Θ. Then we correspond to each row r of Θ the number $Opt_G^\mu(\Theta, r)$ which is equal to the number of rows in Θ which are labeled with decisions different from d.

Let Θ be a nonterminal node of G and all children of Θ have already been treated. Let $r = (b_1, \ldots, b_n)$ be a row of Θ. We correspond the number $Opt_G^\mu(\Theta, r) = \min\{Opt_G^\mu(\Theta(f_i, b_i), r) : f_i \in E_G(\Theta, r)\}$ to the row r in the table Θ, and we set $E_{G^\mu}(\Theta, r) = \{f_i : f_i \in E_G(\Theta, r), Opt_G^\mu(\Theta(f_i, b_i), r) = Opt_G^\mu(\Theta, r)\}$.

From the reasoning before the description of the procedure of optimization relative to the number of misclassifications (first part of Section 4) the next statement follows.

Theorem 2. *For each node Θ of the graph G^μ and for each row r of Θ, the set $Rul_{G^\mu}(\Theta, r)$ is equal to the set $Rul_G^\mu(\Theta, r)$ of all γ-decision rules with the minimum number of misclassifications from the set $Rul_G(\Theta, r)$.*

An example of the directed acyclic graph G^μ obtained after the procedure of optimization relative to the number of misclassifications can be found in [4].

5 Greedy Algorithm

In this section, we present a greedy algorithm for γ-decision rule construction. This algorithm at each iteration chooses an attribute $f_i \in \{f_1, \ldots, f_n\}$ with the minimum index such that uncertainty of corresponding subtable is the minimum. We apply the greedy algorithm sequentially to the table T and each row r of T. As a result, for each row of the decision table T we obtain one decision rule.

Algorithm 1. Greedy algorithm for γ-decision rule construction

---

Require: Decision table T with conditional attributes $f_1, \ldots, f_n$, row $r = (b_1, \ldots, b_n)$ of T, γ that is a nonnegative real number.
Ensure: γ-decision rule for T and r.
  $Q \leftarrow \emptyset$;
  $T' \leftarrow T$;
  while $J(T') > \gamma$ **do**
    select $f_i \in \{f_1, \ldots, f_n\}$ with the minimum index such that $J(T'(f_i, b_i))$ is the minimum;
    $T' \leftarrow T'(f_i, b_i)$;
    $Q \leftarrow Q \cup \{f_i\}$;
  end while
  $\bigwedge_{f_i \in Q}(f_i = b_i) \rightarrow d$ where d is the most common decision for T'.

---

6 Experimental Results

We studied a number of decision tables from the UCI Machine Learning Repository [7]. Some decision tables contain conditional attributes that take unique value for each row. Such attributes were removed. In some tables there were equal rows with, possibly, different decisions. In this case each group of identical rows was replaced with a single row from the group with the most common decision for this group. In some tables there were missing values. Each such value was replaced with the most common value of the corresponding attribute.

Let T be one of these decision tables. We consider for this table the value of $J(T)$ and values of γ from the set $\Gamma(T) = \{\lfloor J(T) \times 0.01 \rfloor, \lfloor J(T) \times 0.1 \rfloor, \lfloor J(T) \times 0.2 \rfloor, \lfloor J(T) \times 0.3 \rfloor, \lfloor J(T) \times 0.5 \rfloor\}$. These parameters can be found in Table 1, where column "Rows" contains number of rows, column "Attr" contains number of conditional attributes, column "$J(T)$" contains difference between the number of rows in a decision table and the number of rows with the most common decision for this decision table, column "γ" contains values from $\Gamma(T)$.

Table 1. Parameters of decision tables and values of γ

Decision table	Rows	Attr	$J(T)$	γ				
				$\lfloor J(T) \times 0.01 \rfloor$	$\lfloor J(T) \times 0.1 \rfloor$	$\lfloor J(T) \times 0.2 \rfloor$	$\lfloor J(T) \times 0.3 \rfloor$	$\lfloor J(T) \times 0.5 \rfloor$
Adult-stretch	16	4	4	0	0	0	1	2
Breast-cancer	266	9	76	0	7	15	22	38
Flags	193	26	141	1	14	28	42	70
Monks1-test	432	6	216	2	21	43	64	108
Monks1-train	124	6	62	0	6	12	18	31
Monks3-test	432	6	204	2	20	40	61	102
Monks3-train	122	6	60	0	6	12	18	30
Mushroom	8124	22	3916	39	391	783	1174	1958
Nursery	12960	8	8640	86	864	1728	2592	4320
Zoo	59	16	40	0	4	8	12	20

We studied the minimum number of misclassifications of irredundant γ-decision rules. Results can be found in Table 2. For each row r of T, we find the minimum number of misclassifications of an irredundant γ-decision rule for T and r. After that, we find for rows of T the minimum number of misclassifications of a decision rule with the minimum number of misclassifications (column "min"), the maximum number of misclassifications of such a rule (column "max"), and the average number of misclassifications of rules with the minimum number of misclassifications – one for each row (column "avg").

Table 2. Minimum number of misclassifications of irredundant γ-decision rules for $\gamma \in \Gamma(T)$

Name of decision table	$\gamma = \lfloor J(T) \times 0.01 \rfloor$			$\gamma = \lfloor J(T) \times 0.1 \rfloor$			$\gamma = \lfloor J(T) \times 0.2 \rfloor$			$\gamma = \lfloor J(T) \times 0.3 \rfloor$			$\gamma = \lfloor J(T) \times 0.5 \rfloor$		
	min	avg	max	min	avg	max	min	avg	max	min	avg	max	min	avg	max
Adult-stretch	0	0.00	0	0	0.00	0	0	0.00	0	0	0.00	0	0	0.00	0
Breast-cancer	0	0.00	0	0	0.09	1	0	0.71	4	0	1.31	5	0	3.19	11
Flags	0	0.00	0	0	0.01	1	0	0.02	1	0	0.09	2	0	0.59	7
Monks1-test	0	0.00	0	0	2.00	4	0	4.50	9	0	6.00	12	0	27.00	36
Monks1-train	0	0.00	0	0	0.21	1	0	0.83	3	0	1.56	6	0	6.36	11
Monks3-test	0	0.00	0	0	0.00	0	0	3.83	9	0	5.67	12	0	19.00	36
Monks3-train	0	0.00	0	0	0.10	2	0	0.60	4	0	1.41	6	1	4.84	12
Mushroom	0	0.00	0	0	0.00	0	0	0.00	0	0	0.00	0	0	1.57	72
Nursery	0	1.21	18	0	65.83	188	0	137.46	390	0	271.20	548	0	878.13	1688
Zoo	0	0.00	0	0	0.00	0	0	0.00	0	0	0.00	0	0	0.20	2

Table 3 presents, for $\gamma \in \Gamma(T)$, the average number of misclassifications of γ-decision rules constructed by the greedy algorithm.

Results presented in Table 2 and Table 3 show that the average value of the minimum number of misclassifications of γ-decision rules is nondecreasing when the value of γ is increasing.

To make comparison of the average number of misclassifications of γ-decision rules constructed by the dynamic programming algorithm and greedy algorithm we consider a relative difference which is equal to:

$$\begin{cases} \frac{GreedyValue - OptimumValue}{OptimumValue} & \text{if } OptimumValue > 0, \\ 0 & \text{if } OptimumValue = 0 \text{ and} \\ & GreedyValue = 0, \\ GreedyValue/OptimumValue & \text{if } OptimumValue = 0 \text{ and} \\ & GreedyValue > 0, \end{cases}$$

Table 3. Average number of misclassifications of γ-decision rules constructed by the greedy algorithm

Decision table	$\gamma = \lfloor J(T) \times 0.01 \rfloor$	$\gamma = \lfloor J(T) \times 0.1 \rfloor$	$\gamma = \lfloor J(T) \times 0.2 \rfloor$	$\gamma = \lfloor J(T) \times 0.3 \rfloor$	$\gamma = \lfloor J(T) \times 0.5 \rfloor$
Adult-stretch	0.00	0.00	0.00	0.25	0.50
Breast-cancer	0.00	3.34	6.87	7.71	7.83
Flags	0.33	3.78	4.27	4.27	4.27
Monks-1-test	0.00	9.00	27.00	27.00	27.00
Monks-1-train	0.00	1.67	8.43	8.43	8.43
Monks-3-test	0.06	3.00	19.00	19.00	19.00
Monks-3-train	0.00	1.37	6.25	6.25	6.25
Mushroom	0.06	16.60	16.60	16.60	16.60
Nursery	19.11	140.82	878.13	878.13	878.13
Zoo	0.00	0.66	1.75	1.75	1.75

where *GreedyValue* is the average number of misclassifications of γ-decision rules constructed by the greedy algorithm (see Table 3), *OptimumValue* is the average number of misclassifications of γ-decision rules constructed by the dynamic programming algorithm (see column "avg" in Table 2).

Table 4 presents, for $\gamma \in \Gamma(T)$, the relative difference of the average number of misclassifications.

Table 4. Comparison of the average number of misclassifications

Decision table	$\gamma = \lfloor J(T) \times 0.01 \rfloor$	$\gamma = \lfloor J(T) \times 0.1 \rfloor$	$\gamma = \lfloor J(T) \times 0.2 \rfloor$	$\gamma = \lfloor J(T) \times 0.3 \rfloor$	$\gamma = \lfloor J(T) \times 0.5 \rfloor$
Adult-stretch	0	0	0	0.25/0	0.5/0
Breast-cancer	0	36.11	8.68	4.89	1.45
Flags	0.33/0	377.00	212.50	46.44	6.24
Monks-1-test	0	3.50	5.00	3.50	0.00
Monks-1-train	0	6.95	9.16	4.40	0.33
Monks-3-test	0.06/0	3/0	3.96	2.35	0.00
Monks-3-train	0	12.70	9.42	3.43	0.29
Mushroom	0.06/0	16.6/0	16.6/0	16.6/0	9.57
Nursery	14.79	1.14	5.39	2.24	0.00
Zoo	0	0.66/0	1.75/0	1.75/0	7.75

Presented results show that the average number of misclassifications of γ-decision rules constructed by the greedy algorithm is usually greater than the average number of misclassifications of γ-decision rules constructed by the dynamic programming algorithm. Often, from $\gamma = \lfloor J(T) \times 0.2 \rfloor$, the relative difference is decreasing when γ is increasing. For $\gamma = \lfloor J(T) \times 0.5 \rfloor$, for data sets "Monks-1-test", "Monks-3-test", and "Nursery", we can observe equal values of the average number of misclassification of γ-decision rules constructed by the dynamic programming algorithm and the greedy algorithm. In this cases, *OptimumValue* > 0 and *GreedyValue* > 0.

We can consider γ as an upper bound on the number of misclassifications of γ-decision rules (see column γ in Table 1). Results presented in Table 2 show that usually the minimum number of misclassifications of a γ-decision rule for each row of the considered decision table is less than γ if $\gamma > 0$ (see column "max" in Table 2). It means that the problem of optimization of γ-decision rules relative to the number of misclassifications is reasonable.

The presented experiments were done using software system Dagger [2]. It is implemented in C++ and uses Pthreads and MPI libraries for managing threads and processes respectively. It runs on a single-processor computer or multiprocessor system with shared memory.

7   Conclusions

We presented a comparison of the number of misclassifications of γ-decision rules using greedy and dynamic programming approaches. Experimental results show that the average number of misclassifications of γ-decision rules constructed by the greedy algorithm is usually greater than the average number of misclassifications of γ-decision rules constructed by the dynamic programming algorithm. The presented results of experiments show also that real value of the minimum number of misclassifications is often less than upper bound on the number of misclassifications given by γ.

References

1. Agrawal, R., Srikant, R.: Fast algorithms for mining association rules in large databases. In: Bocca, J.B., Jarke, M., Zaniolo, C. (eds.) Proceedings of the 20th International Conference on Very Large Data Bases, VLDB 1994, pp. 487–499. Morgan Kaufmann (1994)
2. Alkhalid, A., Amin, T., Chikalov, I., Hussain, S., Moshkov, M., Zielosko, B.: Dagger: A tool for analysis and optimization of decision trees and rules. In: Computational Informatics, Social Factors and New Information Technologies: Hypermedia Perspectives and Avant-Garde Experiences in the Era of Communicability Expansion, pp. 29–39. Blue Herons, Bergamo (2011)
3. Amin, T., Chikalov, I., Moshkov, M., Zielosko, B.: Dynamic programming approach for partial decision rule optimization. Fundam. Inform. 119(3-4), 233–248 (2012)
4. Amin, T., Chikalov, I., Moshkov, M., Zielosko, B.: Optimization of approximate decision rules relative to number of misclassifications. In: Graña, M., Toro, C., Posada, J., Howlett, R.J., Jain, L.C. (eds.) KES. Frontiers in Artificial Intelligence and Applications, vol. 243, pp. 674–683. IOS Press (2012)
5. Amin, T., Chikalov, I., Moshkov, M., Zielosko, B.: Dynamic programming approach for exact decision rule optimization. In: Skowron, A., Suraj, Z. (eds.) Rough Sets and Intelligent Systems - Professor Zdzisław Pawlak in Memoriam. Intelligent Systems Reference Library, vol. 42, pp. 211–228. Springer (2013) (Electronic version available)
6. Amin, T., Chikalov, I., Moshkov, M., Zielosko, B.: Dynamic programming approach to optimization of approximate decision rules. Information Sciences 221, 403–418 (2013) (Electronic version available)
7. Asuncion, A., Newman, D.J.: UCI Machine Learning Repository (2007), http://www.ics.uci.edu/~mlearn/
8. Bazan, J.G., Nguyen, H.S., Nguyen, T.T., Skowron, A., Stepaniuk, J.: Synthesis of decision rules for object classification. In: Orłowska, E. (ed.) Incomplete Information: Rough Set Analysis, pp. 23–57. Physica-Verlag, Heidelberg (1998)

9. Moshkov, M., Piliszczuk, M., Zielosko, B.: Partial Covers, Reducts and Decision Rules in Rough Sets - Theory and Applications. SCI, vol. 145. Springer, Heidelberg (2008)

10. Moshkov, M., Zielosko, B.: Combinatorial Machine Learning - A Rough Set Approach. SCI, vol. 360. Springer, Heidelberg (2011)

11. Nguyen, H.S.: Approximate boolean reasoning: foundations and applications in data mining. In: Peters, J.F., Skowron, A. (eds.) Transactions on Rough Sets V. LNCS, vol. 4100, pp. 334–506. Springer, Heidelberg (2006)

12. Pawlak, Z.: Rough Sets - Theoretical Aspects of Reasoning about Data. Kluwer Academic Publishers, Dordrecht (1991)

13. Skowron, A.: Rough sets in KDD. In: Shi, Z., Faltings, B., Musem, M. (eds.) 16th World Computer Congress, IFIP 2000, Proc. Conf. Intelligent Information Processing, IIP 2000, pp. 1–17. House of Electronic Industry, Beijing (2000)

14. Zielosko, B.: Sequential optimization of γ-decision rules. In: Ganzha, M., Maciaszek, L.A., Paprzycki, M. (eds.) FedCSIS, pp. 339–346 (2012)

Set-Based Detection and Isolation
of Intersampled Delays and Pocket Dropouts
in Networked Control

Nikola Stanković[1,2], Sorin Olaru[1], and Silviu-Iulian Niculescu[2]

[1] SUPÉLEC System Science (E3S), Automatic Control Department,
Gif-sur-Yvette, France
{nikola.stankovic,sorin.olaru}@supelec.fr
[2] Laboratory of Signals and Systems (L2S, UMR CNRS 8506), CNRS-SUPÉLEC,
Gif-sur-Yvette, France
{nikola.stankovic,silviu.niculescu}@lss.supelec.fr

Abstract. This paper focuses on the set-based method for *detection and isolation* of the network-induced *time-delays* and *packet dropouts*. We particularly pay attention to the systems described by a linear discrete-time equation affected by additive disturbances, controlled via *redundant* communication channels prone to intersampled time-delays and packet dropouts. Time-delays and packet dropouts are considered as faults from a classical control theory perspective. We will show that fault detection and isolation can be achieved indirectly through the separation of *positively invariant sets* that correspond to the closed-loop dynamics with *healthy* and *faulty* communication channels. For this purpose, in order to provide a reference signal that uniquely guarantees the fault detection and isolation in real time, we design a *reference governor* using receding horizon technique.

Keywords: Fault detection and isolation, time-delay systems, positively invariant sets.

1 Introduction

High technical demands on safety and performance, for instance in modern aeronautics, require application of redundant sensors and actuators. Unfortunately, numerous examples in practice testify that malfunction in actuation (see Maciejowski and Jones [2003]) and sensing systems (see BEA [2012]) sometimes could end with fatal consequences. Therefore, a great effort has been put in development of control systems which, based on the built-in redundancy, can tolerate malfunctions while maintaining desirable performance (see e.g. Blanke [2003], Seron et al. [2008], Olaru et al. [2010]).

In practical applications, communication between various components in the loop is attained via imperfect channels. Despite the advantages that they brought in control, real networks are prone to undesirable effects such as time-delays (e.g. network access and transmission delays) and packet dropouts due to network

M. Graña et al. (Eds.): KES 2012, LNAI 7828, pp. 51–60, 2013.
© Springer-Verlag Berlin Heidelberg 2013

congestion or transmission errors in physical links (see Hespanha et al. [2007]). Since the presence of time-delays has mostly destabilizing effect, their consideration may be crucial for the overall system behavior (see e.g. Niculescu [2001], Sipahi et al. [2011]). Time-delays and packet dropouts in control over real networks have been exhaustively treated in the literature (see for instance Zhang et al. [2001] and Hespanha et al. [2007]). In most of these works, analysis has been carried out from the robustness point of view i.e. to detect the maximal time-delay in a communication channel (sensor-to-controller and/or controller-to-actuator) which preserves desirable closed-loop behavior. On the other side, an active strategy for constant delay detection and isolation has been recently proposed in Stanković et al. [2012].

A set theoretic approach, based on switching multi-sensor network, with fault detection and isolation (FDI) capabilities in the feedback loop, was proposed by Seron et al. [2008]. It was assumed that sensors were deterministic with additive disturbance while the FDI was achieved through invariant sets separation. The main advantage of such approach was the efficient implementation that required only set membership testing, while all invariant sets were computed offline. The drawback of the method, however, was a priori fixed range of the reference signal. Consequently, an unfortunate choice of the reference may render the detection infeasible. Guided by the generic idea that systems can often manage some modest deviations in the reference signal, Stoican et al. [2012] proposed a solution for this limitation introducing a reference governor in the loop, thus considerably increasing operational range of the method.

Building upon the results of Stoican et al. [2012], Seron et al. [2008], Olaru et al. [2010], and Stanković et al. [2012] in the present note we develop a FDI switching scheme for intersampled time-delays (less than a sampling period) and/or packet dropouts. The usefulness of the proposed method is outlined by an example where even such small time-delays are destructive for the overall performance of the closed-loop dynamics.

The present article is organized as follows. The following section discuss the propose FDI scheme and the problem formulation. The Section 3 outlines results on the construction of the positively invariant sets. Section 4 addresses the fault detection and isolation scenario while Section 5 deals with the reference governor design. Numerical example is provided in the Section 6 and Section 7 presents our conclusions.

Notations

Denote with $\mathbb{R}$, and $\mathbb{Z}$ sets of real and integer numbers, respectively. The closed interval of integers is defined as $\mathbb{Z}_{[a,b]} = \{i : a \leq i \leq b, a, b \in \mathbb{Z}\}$.

Notations $x[k+1]$, $x[k]$ and $x[k-1]$ denote the successor, current and predecessor states, respectively. The Minkowski sum of two sets, $\mathbb{P}$ and $\mathbb{Q}$ is denoted by

$$\mathbb{P} \oplus \mathbb{Q} = \{x : x = p + q, p \in \mathbb{P}, q \in \mathbb{Q}\}.$$

Interior of a set $\$$ is denoted by int($\$$) while the convex hull by conv($\$$). For the p-norm of a vector we use the standard definition, $\|x\|_p = (\sum_{i=1}^{n} |x_i|^p)^{(1/p)}$

where x_i is the ith element of the vector $x \in \mathbb{R}^n$. For a matrix A, we denote its spectral radius by $\rho(A)$.

2   System Description and Problem Formulation

The control scheme considered in the present paper is depicted in the Fig.1. In order to keep the exposition concise, all relevant parts on the scheme are divided by levels, according to their function, and described in the following subsections.

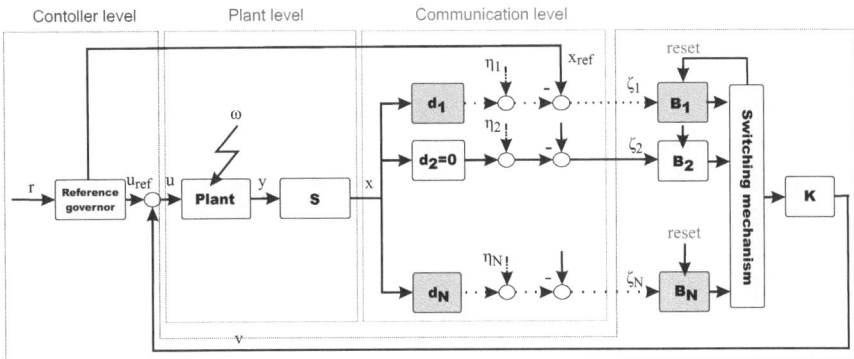

Fig. 1. Multi-sensor fault tolerant control scheme

2.1   Plant Level Description

Consider the problem of regulating the input-output plant model, described by the linear difference equation affected by additive disturbance:

$$y[k+1]+D_1y[k]+\ldots+D_ny[k-n+1] = N_1u[k]+\ldots+N_mu[k-m+1]+\omega[k] \quad (1)$$

where $y[k] \in \mathbb{R}^q$ is the system output, $u[k] \in \mathbb{R}^p$ the system input, $\omega[k] \in \mathbb{R}^q$ bounded disturbance and D_i, $i \in \mathbb{Z}_{[1,n]}$ and N_j, $j \in \mathbb{Z}_{[1,m]}$ are matrices of suitable dimension. We assume that the disturbance $\omega[k]$ is bounded by a polytope with zero in its interior and the sampling is constant and synchronized.

Often in practice, not all of the plant states are available for controller, thus an observer is required. On the other hand, additional dynamics in the loop introduced by observer, could corrupt the fault detection or even make it infeasible.

Nguyen et al. [2011] proposed a method for plant states reconstruction based on the storage of appropriate previous measurements and inputs. Thus, state space representation, for the equation (1), is given as:

$$x[k+1] = Ax[k] + Bu[k] + \omega[k]$$
$$y[k] = Cx[k] \quad (2)$$

where

$$A = \begin{bmatrix} -D_1 & 0_q & \dots & 0_q & I_q \\ -D_n & 0_q & \dots & 0_q & 0_q \\ -D_{n-1} & 0_q & \dots & 0_q & 0_q \\ -D_{n-2} & I_q & \dots & 0_q & 0_q \\ \dots & \dots & & & \\ -D_2 & 0_q & \dots & I_q & 0_q \end{bmatrix}, B = \begin{bmatrix} N_1 \\ N_n \\ N_{n-1} \\ N_{n-2} \\ \vdots \\ N_2 \end{bmatrix}, C = \begin{bmatrix} I_q \\ 0_q \\ 0_q \\ 0_q \\ \vdots \\ 0_q \end{bmatrix}^T, E = \begin{bmatrix} I_q \\ 0_q \\ 0_q \\ 0_q \\ \vdots \\ 0_q \end{bmatrix}.$$

It should be noticed that when $m < n$, then $N_{m+1} = N_{m+1} = \dots = N_n = 0$. We stress here that the state-space representation (2) is certainly minimal only for SISO case, while otherwise is not necessarily.

Using stored certain plant information, written in the vector form as

$$X[k] = \begin{bmatrix} y[k]^T & \dots & y[k-n+1]^T & u[k-1]^T & \dots & u[k-n+1]^T \end{bmatrix}^T, \quad (3)$$

the state vector could be obtained from the following relation:

$$x[k] = TX[k] \quad (4)$$

where

$$T = [T_1 \; T_2], T_1 = \begin{bmatrix} I_q & 0_q & 0_q & \dots & 0_q \\ 0_q & -D_n & 0_q & \dots & 0_q \\ 0_q & -D_{n-1} & -D_n & \dots & 0_q \\ \dots & \dots & \dots & \dots & \dots \\ 0_q & -D_2 & -D_3 & \dots & -D_n \end{bmatrix}, T_2 = \begin{bmatrix} 0_{q \times p} & 0_{q \times p} & 0_{q \times p} & \dots & 0_{q \times p} \\ N_n & 0_{q \times p} & 0_{q \times p} & \dots & 0_{q \times p} \\ N_{n-1} & N_n & 0_{q \times p} & \dots & 0_{q \times p} \\ \dots & \dots & \dots & \dots & \dots \\ N_2 & N_3 & N_4 & \dots & N_n \end{bmatrix}.$$

In the present article we consider scenario where the 'raw' sensor measurements are already locally processed, according to the equation (4) (block S in Fig.1), and send over the imperfect, redundant channels at each sampling period. For more details on outlined state reconstruction see (Nguyen et al. [2011]). Without loss of generality and in order to simplify the exposure as much as possible, in this note we consider the SISO plant model.

The control objective is to design a closed-loop control scheme such that the plant model (2) tracks a reference signal $x_{ref}[k] \in \mathbb{R}^n$ which obeys the nominal dynamics:

$$x_{ref}[k+1] = Ax_{ref}[k] + Bu_{ref}[k]. \quad (5)$$

In order to assure delay detection and isolation in the loop, the pair $x_{ref}[k], u_{ref}[k]$ are provided by a reference governor such that the given constraints are respected and an ideal reference $r[k]$ is followed as closely as possible (see Section 5).

2.2 Communication Level Description

Processed plant output, $x[k]$, is transmitted through N redundant communication channels at each constant sampling period, here for simplicity considered as

$h = 1$. All channels are subjected to delays $0 \leq d_j < 1$, possible packet dropouts and bounded noise η_j, where $j \in \mathbb{Z}_{[1,N]}$. It is assumed that the noise in channels is bounded by a polytope with 0 in its interior.

2.3 Controller Level Description

At each sampling period, information sent through the communication level, possibly corrupted during transmission (see Sect. 2.2), are acquired by buffers B_j. We assume that all buffers have enough memory to store information from one sampling period.

From the classical fault detection and isolation point of view (Blanke [2003]), a signal called residual, sensitive to fault occurrence and with manageable dependence on the disturbances, can be defined for fault detection.

The data from buffers are employed in residual signals generation. Based on the membership testing (in more details exposed in Sect. 4), each sampling period switching mechanism selects one healthy residual signal which is employed in control action, while all the others buffers, not necessarily faulty, are reset to that value.

Let denote with ζ_n the nominal residual signal:

$$\zeta_{nj}[k] = x[k] + \eta_j - x_{ref}[k] = z[k] + \eta_j \tag{6}$$

where $z[k] = x[k] - x_{ref}[k]$ is the plant tracking error.

In situations when feedback information is lost and/or delayed in a communication channel, corresponding buffer keeps the data from the previous sampling period. Therefore, the faulty residual signal is defined in the following manner:

$$\zeta_j[k] = x[k-1] + \eta_j - x_{ref}[k] = z[k-1] + \eta_j + (x_{ref}[k-1] - x_{ref}[k]). \tag{7}$$

The fault that we consider here schematically can be represented as:

$$\zeta_{n_i}[k] \xrightarrow{FAULT} \zeta_i[k]. \tag{8}$$

Under the assumption that the pair (A, B) from (2) is stabilizable, one could obtain the feedback gain K employing one of well known methods from the control theory (LQR, LQG etc.). The details of such a synthesis procedure are omitted here.

3 Invariant Set Construction

Based on the set membership testing of residual signals, at each sampling period, the switching mechanism selects one feedback channel suitable for control. Requirement to guarantee the confinement of residuals in certain regions of their associated space, leads us to introduction of invariant sets with respect to the respective dynamics. By definition, invariant sets are able to preserve the states of particular dynamics for any initial data from that set. In this section, we address the problem of computing suitable invariant sets for fault detection scheme.

In order to assure unique fault detection, invariant sets for the healthy and faulty residuals need to be disjoint. Furthermore, the interest to have more freedom in choosing the reference signal, and subsequently, less conservative results, impose the use of the minimal robust positively invariant (mRPI) sets.

Definition 1. *The set $\$ \subseteq \mathbb{R}^n$ is said to be robust positively invariant (RPI) with respect to the system (2) if for all $x[0] \in \$$ and for all $\omega[k] \in \Omega$, all solutions satisfy $x[k] \in \$$ for $\forall k \in \mathbb{Z}_{[0,\infty]}$ i.e. $A\$ \oplus \Omega \subseteq \$$. If the set $\$$ is contained in any closed RPI set, then it is mRPI set and it is unique, compact and contains the origin if Ω does.*

The efficient computation of an invariant approximation of the mRPI set was proposed by (Raković et al. [2005]) and is outlined in the following theorem:

Theorem 1. *Consider the system (2), controlled by $u[k] = -Kx[k]$ such that $\rho(A - BK) < 1$ and $w[k] \in \Omega$, $0 \in \Omega$. Then there exist a finite integer $s > 0$ and a real scalar $0 < \alpha < 1$ such that*

$$(A - BK)^s \Omega \subseteq \alpha\Omega.$$

The compact and convex invariant approximation of the mRPI set is given as:

$$\$(s, \alpha) = (1 - \alpha)^{-1}\$(s)$$

where $\$(s) = \bigoplus_{i=0}^{s-1}(A - BK)^i \Omega$, $\$(0) = \{0\}$.

For further details on the minimal robust positively invariant set construction see (Raković et al. [2005]).

4   Fault Detection and Isolation

The aim of the present work is to propose manageable realization for control, through redundant communication channels, in the presence of time-delays or packets dropouts.

Robustness issues are not addressed and they will be the scope of our future research. Thus, for the proper functioning of the pattern, we assume that at any time sequence at least one of the communication channels carry the proper plant information, possibly affected only by additive disturbances.

The purpose of the switching mechanism is to prevent closing the control loop with 'obsolete' plant information. The plant tracking error for the (2) is given as:

$$z[k + 1] = (A - BK)z[k] + \psi_j[k], \tag{9}$$

where $\psi_j[k] = \omega[k] - BK\eta_j$ is absolute additive disturbance, also bounded by a polytope, denoted by $\mathbb{N}_j \subset \mathbb{R}^n$, since the same hold for $\omega[k]$ and η_j. For the simplicity, we immerse the polytopic bounds for all channels in one polytope defined as $\mathbb{N} = \text{conv}(\mathbb{N}_j), j \in \mathbb{Z}_{[1,N]}$.

Let denote by $\$(s,\alpha)$ mRPI invariant set for (9), computed according the Theorem 1.

Since the bounds for disturbances in the channels are predefined, we are able to calculate the invariant regions where the healthy (R_n) and the faulty (R) residuals are settled down:

$$R_n = \$(s,\alpha) \oplus \mathbb{N}$$
$$R = \$(s,\alpha) \oplus \mathbb{N} \oplus (x_{ref}[k-1] - x_{ref}[k]), \tag{10}$$

where we used the fact that $z[k-1] \in \$(s,\alpha)$.

If the following relation holds

$$R_n \cap R = \emptyset, \tag{11}$$

then, by validating that $\zeta_{ni}[k] \in R_n$, we can affirm that the ith loop operate in desirable way and it can be used by the controller. It should be noticed that fault detection depends on the reference signal i.e. on the term $(x_{ref}[k-1] - x_{ref}[k])$. Therefore, an appropriate selection of the reference, results in disjoint invariant regions for healthy and faulty channels, thus guaranteeing detection in real time. Formally, the separation is guaranteed if the following condition holds:

$$(x_{ref}[k-1] - x_{ref}[k]) \notin (\$(s,\alpha) \oplus \mathbb{N}) \oplus (-((\$(s,\alpha) \oplus \mathbb{N}))). \tag{12}$$

5 Reference Governor Design

In the previous section we obtained the condition on the reference signal (12) that guarantees the invariant sets separation for healthy and faulty behavior. Define the set of admissible reference values as:

$$\mathbb{H} = \{x_{ref}[k] : (\$(s,\alpha) \oplus \mathbb{N}) \cap (\$(s,\alpha) \oplus \mathbb{N} \oplus (x_{ref}[k-1] - x_{ref}[k])) = \emptyset\}. \tag{13}$$

Since we constrain the state and input references to take values only from their admissible set, we may no longer follow the desired trajectory. Consequently, a pair of input/state reference will be sought in order to satisfy the dynamics (5) and to minimize the mismatch between an ideal and real trajectories, always respecting the imposed constraints in (13). To this end, we propose the use of a reference governor, implemented though receding horizon techniques.

The set $\mathbb{H}$ (13) can be rewritten in a more readable form as:

$$\mathbb{H} = \{(x_{ref}[k], u_{ref}[k]) : (x_{ref}[k-1] - x_{ref}[k]) \notin (\$(s,\alpha) \oplus \mathbb{N}) \oplus (-((\$(s,\alpha) \oplus \mathbb{N}))). \tag{14}$$

The feedforward action u_{ref} is provided by the reference governor, which has to choose a feasible reference signal and, at the same time, follow an desired reference as close as possible. This problem can be cast as the optimization of a cost function under constraints, and it is solved in a model predictive control framework:

$$u^* = \underset{u_{ref}[k,k+\sigma-1]}{\arg\min} \left(\sum_{i=1}^{\sigma} \|(r[k+i] - x_{ref}[k+i])\|_{Q_r}^2 + \sum_{i=0}^{\sigma-1} \|(u_{ref}[k+i])\|_{P_r}^2 \right) \tag{15}$$

subject to:

$$x_{ref}[k+1] = Ax_{ref}[k] + Bu_{ref}[k]$$
$$(x_{ref}[k-1] - x_{ref}[k]) \in \mathbb{H},$$
(16)

where $\sigma \in \mathbb{Z}_{(0,\infty)}$ is a prediction horizon, $Q_r \succ 0$ and $P_r \succ 0$ are weighting matrices and $u_{ref}[k]$ is taken as the first element of the sequence u^*.

Remark 1. One can observe that the set $\mathbb{H}$ is non-convex. As a consequence, the optimization problem has to be solved over a non-convex set which imposes the use of mixed-integer techniques. For more details we refer to (Prodan et al. [2011]).

6 Ilustrative Example

Consider the linear time invariant plant model with two feedback communication channels:

$$x[k+1] = \begin{bmatrix} 0.9 & -0.34 \\ 0.84 & 0.7 \end{bmatrix} x[k] + \begin{bmatrix} 1 \\ 1 \end{bmatrix} u[k] + \omega[k]$$
(17)

where $\omega[k] \in \Omega = \{x : \|x\|_\infty \le 0.1\}$.

Assume that the feedback channels are affected by random measurement noise $\eta_j \in \mathbb{N}$ and, where $\mathbb{N} = \{x : \|x\|_\infty \le 0.1\}$, $j = 1, 2$. LQR gain matrix for this case is $K = \begin{bmatrix} 0.6841 & 0.1786 \end{bmatrix}$.

Initial reference signal r and the reference signal obtained from the reference governor x_{ref} are depicted in the Fig.2. Real-time simulation i.e. tracking error, depicted in the Fig.3, exhibit the behavior of the considered network. From the simulation we can notice that the delay is efficiently avoided as long as there is at least one uncorrupted channel.

The outlined results shows the simplicity and, at the same time, efficient detection and isolation of corrupted communication channels by an intersampled time-delay or a packet dropout.

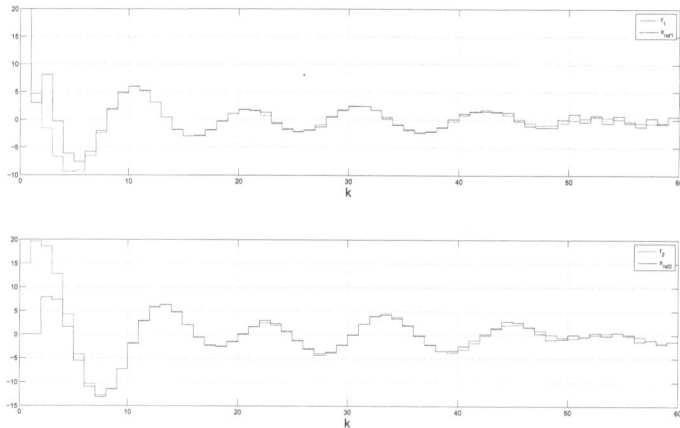

Fig. 2. Ideal reference r and the reference governor output x_{ref}

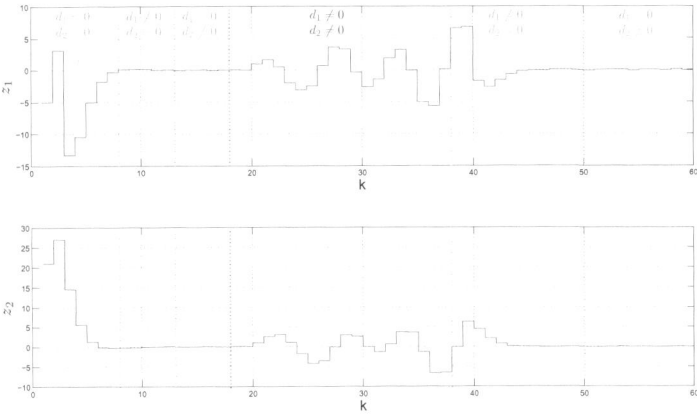

Fig. 3. Plant tracking error

7   Conclusions

The paper has presented a fault tolerant control scheme based on a reconfigurable control action, where delays or communication packet dropouts in the redundant feedback channels are considered as faults. The reference, followed by the system, was obtained through the reference governor which employs a receding horizon technique in order to determine a reference which guarantees correct fault detection in real time. The proposed scheme could be employed in a reference signal construction in order to, indirectly, excluding defective channels from being utilized in the control action. The proposed method is computationally efficient for the online implementation since it is carried out only by set membership testing, while the invariant sets computation can be done offline.

References

[2012]   BEA (Bureau d'enquêtes et d'analyses pour la sécurité de l'aviation civile): Final report on the accident on 1st June 2009.
http://www.bea.aero/en/enquetes/flight.af.447/rapport.final.en.php, (2012)

[2003]   Blanke, M.: Diagnosis and fault-tolerant control. Springer Verlag (2003)

[2007]   Hespanha, J.P., Naghshtabrizi, P. and Xu, Y.: A survey of recent results in networked control systems. Proceedings of the IEEE, 95, 138–162 (2007)

[2003]   Maciejowski, J.M. and Jones, C.N.: MPC fault-tolerant flight control case study: Flight 1862. IFAC Safeprocess Conference, (2003)

[2011]   Nguyen H.N., Olaru S., Gutman P.O. and Hovd M.,: Robust output feedback model predictive control. 8th International Conference on Informatics in Control, Automation and Robotics, (2011)

[2001]   Niculescu, S.I.: Delay effects on stability: A robust control approach. Springer Verlag 269, (2001)

[2010]   Olaru, S., De Doná, J.A., Seron, MM and Stoican, F.: Positive invariant sets for fault tolerant multisensor control schemes. International Journal of Control 83, 2622–2640 (2010)

[2011] Prodan, I., Stoican, F., Olaru, S. and Niculescu, S.I.: Enhancements on the Hyperplanes Arrangements in Mixed-Integer Techniques. Journal of Optimization Theory and Applications, 154(2), 549–572 (2012)

[2005] Rakovic, S.V., Kerrigan, E.C., Kouramas, K.I. and Mayne, D.Q.: Invariant approximations of the minimal robust positively invariant set. IEEE Transactions on Automatic Control, 50, 406–410 (2005)

[2008] Seron, M.M., Zhuo, X.W., De Doná, J.A. and Martínez, J.J.: Multisensor switching control strategy with fault tolerance guarantees. Automatica 44, 88–97 (2008)

[2011] Sipahi, R., Niculescu, S.I., Abdallah, C.T., Michiels, W. and Gu, K.: Stability and stabilization of systems with time delay. IEEE Control Systems Magazine, 31, 38–65 (2011)

[2012] Stanković, N., Stoican, F., Olaru, S. and Niculescu, S.I.: Reference governor design with guarantees of detection for delay variation. 10th IFAC Workshop on Time Delay Systems, Time Delay Systems 10(1), 67–72 (2012)

[2012] Stoican, F., Olaru, S., De Doná, J.A. and Seron, M.M.: Reference governor design for tracking problems with fault detection guarantees. Journal of Process Control, available online at http://www.sciencedirect.com/science/article/pii/S095915241200039X (2012)

[2001] Zhang, W., Branicky, M.S. and Phillips, S.M.: Stability of networked control systems. IEEE Control Systems Magazine, 21, 84–99 (2001)

Analysis and Synthesis of the System for Processing of Sign Language Gestures and Translatation of Mimic Subcode in Communication with Deaf People

Wojciech Koziol[1], Hubert Wojtowicz[1], Kazimierz Sikora[2], and Wieslaw Wajs[3]

[1] University of Rzeszow, Faculty of Mathematics and Nature,
Institute of Computer Science, Rzeszow, Poland
[2] Jagiellonian University, Faculty of Polish Language,
Chair of History of Language and Dialectology, Cracow, Poland
[3] AGH University of Science and Technology, Faculty of Electrical Engineering,
Institute of Automatics, Cracow, Poland

Abstract. In the paper a design and principle of operation of the system facilitating communication between hearing and deaf people is presented. The system has a modular architecture and consists of main application, translation server and two complementary databases. The main application is responsible for interaction with the user and visualization of the sign language gestures. The translation server carries out translation of the text written in the Polish language to the appropriate messages of the sign language. The translation server is composed of facts database and translation rules implemented in the Prolog language. The facts database contains the set of the lexemes and their inflected forms with a description of the semantics of units. The translation rules carry out identification and analysis of basic structures of the Polish language sentence. These basic structures are related to the sentence creation function of the verb predicate. On the basis of this analysis equivalent translation of text into the sign language is realized. Translated text in the form of metadata is passed to the main application, where it is translated into the appropriate gestures of the sign language and face mimicry. The gestures in the form of 3d vectors and face mimicry are stored in the main database as binary objects. The authors intend to apply the translation system in various public institutions like hospitals, clinics, post offices, schools and offices.

1   Introduction

The research on issues related to the translation of texts written in the Polish language into the Polish sign language and vice versa, showed the complexity of the problem. The lack of relevant publications fully describing the grammar of the Polish sign language, scarcity of the sources of information on the unified gestures and internal divisions among people using sign language into the supporters of Polish Sign Language (PSL) and System of Sign Language (SSL)

M. Graña et al. (Eds.): KES 2012, LNAI 7828, pp. 61–70, 2013.
© Springer-Verlag Berlin Heidelberg 2013

are the main difficulties encountered during the implementation of automatic text translation between these two languages. This is a complex issue both from the point of view of linguistic as well as that of computer science. The implementation of such a system requires in-depth knowledge of general linguistics, semantics and syntax and the application of appropriate tools and solutions provided by computer science. For the language (phonic) communication the most important are the compound signs and propositional structures. These structures contain apart from verbal components and propositional contents, also intentional properties including obligatory predicate of modality: affirmative, interrogative and imperative. The utterance in the Polish language is characterized by the presence of indicators of the pragmatic functions like commands, advices, requests, wishes, threats etc. In a broader sense the sentence is also characterized by the indicators of the content expressed indirectly, which may include implicatures, presuppositions and elements of actual fragmentation of communication, in other words the thematic-rematic structure expressed to a large degree by accentuation and intonation. Preliminary research studies have established that the seemingly simple messages in sign language are however characterized by a high degree of complexity. Each sign language gesture has its own content structure interpreted by the face mimicry. The analysis of the Polish language expressions and translation into the messages of the sign language is achieved through the application of Prolog language. This language is dedicated to solving problems concerning symbolic processing. Transfer of information in the sign language is achieved by visualization of the gestures together with face mimicry in the 3D technology. Ideograms are acquired by using a motion capture system recording spatial data in function of time. Another complex task is joining of the individual ideographic characters into the compound messages. It requires solving the problem of transitions between particular gestures, which cause collisions of arms and hands of the avatar. Transitions of this type require an earlier intervention through the detection and appropriate handling of collisions. The realization of the system translating messages of the sign language into the Polish language text seems to be even more complex issue. The analysis of the research carried out domestically leads to the conclusion that the system, which solves this issue to the sufficient degree, has not been realized until now. This problem is researched using complex computational intelligence and spatial image analysis methods. The analysis and identification of sign language gestures is realized on the basis of an extracted set of distinctive features. A method of gesture analysis used by the authors takes advantage of a special outfit allowing recording of the spatial movement. The data acquired using such an outfit is subject to analysis, which allows identification of the gestures on the basis of their characteristic features. This method of gestures' recognition is however much more troublesome, comparing to the approaches not using additional outfit, because of the strict requirements of matching the outfit to the figure of the signing person. The results of the research conducted worldwide clearly show high degree of complexity of the issues addressed in the author's work. Another very important issue exists neglected until now. It concerns structuring of the recognized gestures (in the form of text equivalents) on the level of syntactic organization and linear system. Deep analysis of the order of the previously recognized gestures should be taken into account including syntactic (on the level of sentence structuring) and semantic connectivity. This analysis should allow equivalent translation of the sequence of gestures into the statements in the Polish language. The solution of this problem requires development of appropriate algorithms in the Prolog language.

2   The Aim of the Work

The primary objective at the current stage of work is the development of an information system taking advantage of 3D technology and multimedia movie sequences technology. Its primary component is a dictionary. Sign language gestures are visualized in two alternative forms, which are film sequences and 3D animations. The unification of many of the sings used in the translation was proposed in the project. An open dictionary of the sign language was implemented, which can serve as the basis for the adoption of a common gestures lexicon by interested sign language communities. The project is of interdisciplinary character and combines the achievements of modern knowledge about language in the area of morpho-syntax, syntactic and sentence semantics, with computer science tools used for the analysis of language. The system uses two independent processes: the main application and a translation server. These processes work together in a client-server architecture using the TCP/IP protocol. The main application is responsible for the interaction with users. A hearing user inputs a text in the Polish language which is redirected to the translation server through the main application. The translation module is implemented in the Prolog language. The program includes various individual complementary modules conducting the analysis of sentences and phrases, determining their lexical representation in the facts database, and their equivalence at the appropriate structural level (in its final version the system will analyze compound sentences - the current version analyses only isolated sentences). Sentence analysis is carried out by isolating individual phrases, and then identifying particular parts of the sentence. Analyzing rules require collecting an appropriate knowledge base in the form of the facts of the Prolog language. This knowledge base contains grammar -inflectional paradigms of the Polish language and semantics. The list of the lexemes for the facts database is defined on the basis of the "Illustrated Basic Dictionary of the Polish Language" [2] and supplemented with vocabulary from various dictionaries and texts concerning sign language [1] [3] [5] [6] [7] [8] [9]. The facts database comprises about 8 thousands lexemes. The introduction of semantic markers into the facts database allows the analysis of the deep structures of the sentence. The set of the semantic markers, together with the schemes of semantic collocation for the particular verbs, is adopted from "Dictionary of Syntactic Generative Verbs of the Polish Language" [4]. The application of grammar and semantic collocation rules allows the implementation of a set of rules for the analysis of the sentence in the Polish language and the synthesis of this sentence into the form of sign language gestures.

3   Assumptions for the Translation of Polish Language Sentences into Sets of Features and Sets of Rules

The main assumption of the translation system is the establishment of the equivalence principle on the level of basic cognitive structures. These structures are represented by elementary syntactic structures (predicative-argumentative structures) i.e.

- unary predicate - Tomasz pracuje. (*eng. Tom is working.*)
- binary predicate - Adam pisze wypracowanie. (*eng. Adam is writing an essay.*)
- ternary predicate - Jan daje zonie kwiaty. (*eng. John is giving flowers to his wife.*)
- quaternary predicate - Wojtek czesze corce wlosy grzebieniem. (*eng. Tom is combing his daughter's hair.*)

In creating these structures the main element is verb connotation consisting in opening positions for syntactic elements such as the subject, direct object, indirect object, and, in some cases, adverbials of manner, place and time, which usually remain optional components of the sentence. Therefore it can be said, that the verbs impose the scheme of the sentence and determine syntactic positions that can appear in their surroundings. Identification of the sentence's scheme, for which the information is embedded in the verb is an essential step ordering the analysis process. Referring to the description of the state of affairs presented in the specialized scientific literature it can be assumed that the number of these strictures and schemes is limited. Therefore they can be formalized and implemented using the Prolog language. Previous attempts to create a computer system translating from the Polish language into the sign language revealed that the basic difficulty was lack of formalized grammar description of this language. It seems reasonable to assume, that to formalize the equivalence rules they should be based on the advices received from people using sign language. It should be noted that these people often don't have sufficient knowledge in this field. Another problematic issue is the lexical resource for the Polish sign language counting barely few thousand ideograms. For comparison in the Polish natural language full lexical scope is counted in hundreds of thousands ideograms. A significant difficulty arising before the authors of the project is the fact, that currently deaf people don't use pure ideographic sign language. Mixed systems (pidgins) are often in use close structurally to the phonic language. Large differences can be observed between regions, environments and generations. During the system implementation the most neutral and simultaneously the most widespread dialect of the Polish sign language was assumed as the reference point. The establishment of equivalence on the level of lexical units requires selecting base vocabulary of the Polish language. "Illustrated Basic Dictionary of the Polish Language" [2] was adopted for this purpose. Repertoire of the sign language gestures was developed on the basis of available dictionaries of the Polish signed and sign languages i.e. "Sign Language Dictionary" [1] and other book sources. For the visualization of the sign language gestures appropriate 3D model of the virtual avatar was created. Control of the avatar is realized using skeletal animation technique. To improve the naturalness and the clarity of communication algorithms enabling controlling of the face mimicry of the model were implemented. Transfer of the content in the sign language can take place on multiple paths with the use of the following channels: ideographic, phonic, face mimicry and text. Utilization of multiple, and to a large degree complementary, communication paths allows to minimize the results of inevitable imperfections of the translation and misunderstandings.

4   Design of the Translation Module

The architecture of the translation module proposed in the project is designed using the Prolog language. The system consists of four fundamental parts:

1. Facts database containing vocabulary items belonging to different parts of speech;
2. Rules database containing rules classifying parts of speech and auxiliary functions;
3. Rules of propositional templates database containing rules concerning the syntactic and semantic analysis of the sentences;
4. Main rule in the form of overloads which applies templates of syntactic analysis to particular texts and makes the translation; the main rule serves as a determination and diagnostic rule in regard to other rules.

The grammatical and semantic facts database contains data about the morphological and syntactic features of the lexemes and their semantics. Building the facts database required the use of www interfaces with MySQL database. This preliminary form of the data was later translated to the facts database in the Prolog language. It consists of: verbs, nouns, adjectives, numerals, adverbs and other, semantically dependent parts of speech (conjunctions, particles, prepositions). Each independent lexeme in the database is defined by a set of features - formal and grammatical (inflexion, syntax) and, if required for the phrase construction, semantic (described by a label which classifies its meaning). The assumption of the direction of the translation - written language → rules database of the expert system - determined the form and scope of linguistic analysis (see above translation assumptions). This information characterizes the constitutive clause of the sentence, represented by the verb, and belongs to its vocabulary features. Analyzing any sentence created in the Polish language, one is able to discern its elements reflecting elementary syntactic functions (positions), determined by the scheme. These elements include: subject (S), direct object 1 (O1), indirect object 2, 3, 4 (O2, O3, O4), predicative, adverbial (A1, A2, A3) and attribute.

The mandatory occurrence of the parts of the sentence mentioned above, connotated by the verb, is determined strictly by its connotation scheme. The status of the attribute in the structure of the sentence is special because it is a modifier of the noun in the nominal phrase.

The design of the application is shown in the block diagram indicating the functional relationships of its main components (Fig. 1).

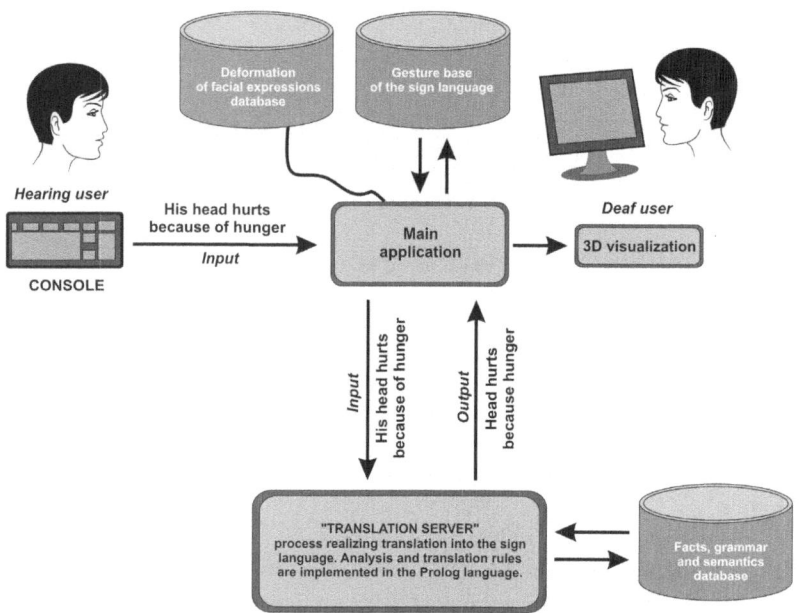

Fig. 1. Scheme of the translation module design

The mechanics of the system can be described as follows. Input data in the form of written language (Input List) is subjected to analytic procedures consisting of several steps:

1. In the first step the identification of the main verbal predicate of the sentence is made together with the identification of the sentence scheme constituting its integral feature.
2. After the identification of the scheme its acquisition phase in the form of analytic conduct rule is carried out. A strictly correlated decision procedure is then started. Its first component is the identification of mandatory positions, syntactic functions (clauses): subject, direct object, indirect object or mandatory object in the sentence structure of the adverbial.
3. The next step of the analytic procedure is the identification of non-mandatory clauses of the sentence.
4. The final step of the analytic procedure consists in the recognition of the structures of the nominal groups and attributes.
5. Ordered facts of the source information are subjected to the process of equivalence determination at the level of sign language' units and structures.

Created translation module works on the basis of the gathered knowledge base. Currently the database contains 3610 lexical forms of nouns, 3556 forms of adjectives, and 1088 forms of verbs together with morphological elements and semantic descriptions. The database contains also forms of pronouns, numerals, adverbs and conjunctions. Equipping word forms with semantic markers allows the analysis of the deep structures of the sentence. It allows intelligent contextual analysis of sentences. Frame rules of the semantic syntax, which means semantic valence of verbs, help in the reduction of polysemia, and simultaneously impose both the syntax model, which is subject to interpretation, and collocation rules (connotation). This kind of information tool combines elements of artificial intelligence and can be used not only in analytic tasks but also to generate text. The system supports approximately 40 sentence schemes. These schemes allow the analysis of affirmative, negative and interrogative single sentences. The module is capable of sentence translation carried out in a sequential manner.

5   Organization of the Communication System between Hearing and Deaf People – Proposed Approach

The text analysis phase aims to provide maximum amount of information about how the translated sentences should be communicated in the sign language, i.e. sentences modalized affirmatively, negatively and interrogatively fundamentally differ in regard to the facial mimicry and it's the description of the facial mimicry which is a very important part of the sign language grammar. Identification of the particular parts of the sentence leads to the synthesis phase. In this phase combining of sign language messages is realized using rules defining syntax of this language - determination of equivalence. Output message should contain information about the behavior of 3D avatar, i.e. ideogram - translation server provides information about the unique number, which is the ID of the animation stored in the database in the form of temporal and spatial 3D motion vectors. Translation server also provides information about the facial mimicry in order to express information during the speech animation process of the avatar. Expressiveness of the facial expressions is very important for the deaf people because it informs about the moods and intentions of the interlocutor and is also an element of the sign language grammar. Particular attention should be paid to the problem of clear illustration of the mouth mimicry of the letters (and phones) in particular words and to the problem of smooth transitions between letters in the spoken word. Deaf people often support the communication through lip-reading, especially in contacts with hearing people. Enrichment of communication with speech synthesis and displaying of the text during animation constitutes additional element supporting understanding of the content communicated in the sign language. Communication paths synchronized (and largely complementary) in this manner should be in our opinion sufficient for the understanding of the statements communicated in the sign language. Output message is a list containing identifiers of the gestures stored in the database and information concerning facial mimicry for the particular parts of sentence. The list combined in the synthesis phase is sent to the main application, which performs visualization of the data. To accomplish this task main application connects to the database containing gestures of the sign language and retrieves binary objects according to the order of the list of identifiers created by the translation server.

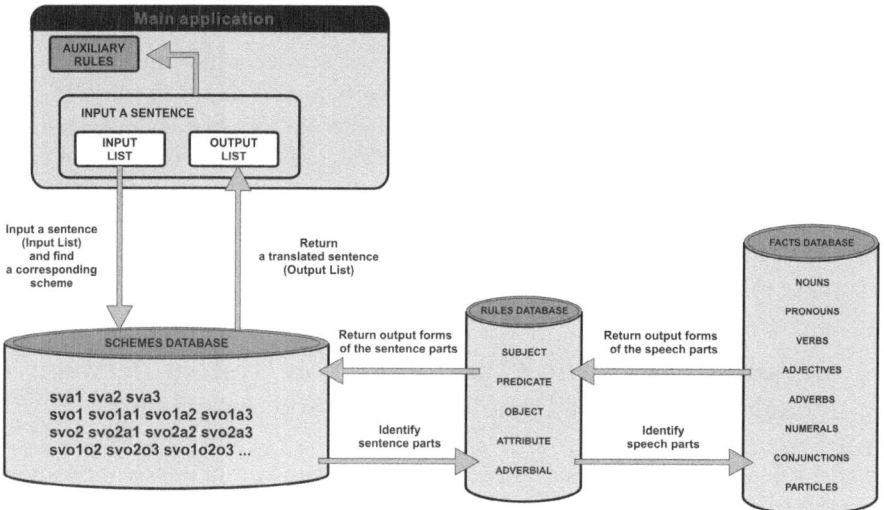

Fig. 2. Flowchart of the one way communication process between a hearing person and a deaf person

These objects contain information about the rotations of the bones of the model, which is arranged in the form of animation frames.

Skeleton and facial mimicry animations are realized in parallel. Facial mimicry animation is interpolated between consecutive key frames. These frames are retrieved from the database containing facial mimicry deformations according to the output list defined by the translation server. Essentially rendering of consecutive frames relies on the combination of two component animations: facial mimicry and rotations of the bones of the model. The synchronization of these two transmission paths is very important for the deaf people because face expressions provide a significant part of the grammar information (in the categorical sense) in the Polish sign language. Flowchart showing the operation of the described system is presented in Fig. 2.

6   Avatar Construction and Acquisition of Sign Language Gestures

The visualization of sign language gestures with face mimicry required construction of the appropriate 3D model - avatar. Created model consists of 11 thousand vertices divided into 93 groups. Each of the vertices' groups is assigned to the particular bone. This assignment enables control of the group of vertices with bones. Application of this big number of bones enables precise representation of human body and limbs movement in 3D space. Creation of the 3D model required appropriate modeling of the mesh. The most important parts of the avatar are head and hands, which require the greatest detail. In order to achieve greater naturalness and better clarity of sign language communication, three types of textures were created for the model: colour textures, specular textures and normal mapping textures. Rendering of bump and reflection maps is realized using Vertex Shader and Pixel Shader technologies. To increase the number of frames displayed per second bump and reflection maps are rendered only for the face and hands. Acquisition of sign language gestures required application of the motion capture system registering locations of the markers using infra red radiation. In the research BTS Bioengineering SMART-E Motion Captures System was used for the capture of sign language gestures. System is equipped with six cameras working in the 120 Hz frequency. For the animation acquisition 30 markers were used placed in the appropriate locations of the body of the animator. Motion capture requires high precision in regard to the placement of the markers on the body of the person. Particular markers shouldn't change their position during following capture sessions. To achieve this accuracy an outfit was created using special rubbers, which ensure tight adherence of the outfit to the body. The outfit was equipped with appropriate attachments for the markers, which enable rapid installation of the markers. The consequence of the motion capture process is animation correction. Distortions result from the erroneous data acquisition during the execution of some of the gestures by the animator. Correction of acquired animations can be done in manual or automatic manner. Automatic correction is realized through application of appropriate interpolation algorithms in order to fill in missing data. Close placement of the hands' and in particular fingers' markers is a limitation of the system. It hinders correct data acquisition and forces separate modeling of the hand movement for the visualization of the gestures. Finite number of possible hand arrangements in sign language gestures enables realization of the procedure of separate animation of the hands in regard to the animation of other body parts. Two possibilities exist for the hand arrangements identification during the animation process. First of them is outfitting the animator with two virtual gloves in order to register hand arrangements during the animation process. Application of computational intelligence algorithms, i.e. MLP neural networks, for the interpretation of this data allows identification of hand movement sequences representing correct execution of gestures and reject random and transitional sequences. Animation data has both spatial and temporal dimension. These dimensions are synchronized with each other and as such are identified by the neural network. Neural networks' task is identification of initial and final hand arrangement. On the basis of this recognition interpolation of transitional states between hand arrangements for particular gestures is realized. Another possibility is manual establishment of hands arrangements. In the sign language fifty different hand arrangements exist. All of them are defined in the unique manner and are stored in the database in the form of binary objects. These objects contain complete information about flexion of fingers expressed by the rotation angles in 3D space.

7   Summary

In the paper an original design of the system enabling translation of the statements in the Polish language into the sign language is described. Results of the research on this issue have reached a high level of advancement and encourage further work on this project. From the technical standpoint the system enables analysis and translation of the sentences corpora and visualization of the gestures in DirectX technology. Considering the technical applications an implementation of appropriate software in Silverlight technology is planned, which provides the functionality of text translation through the www service. Deployment of a comprehensive system is also planned, which aim is supporting the communication between hearing and deaf people. The intent of the authors is implementation of the system in several versions. A standalone application designed for people interested in learning the sign language is one of the planned implementations. For the laymen, just getting to know the sign language, the application may become a broad knowledge source about the sign language and the forms of communications with deaf people. Another version of the system is implemented as a kiosk, which is a device allowing translation of the text in the Polish language in the sign language in particular life situations. These devices can be utilized by various public institutions like health care facilities, post offices, schools etc.

References

1. Hendzel, J.K.: Sign Language Dictionary, Offer, Olsztyn (1995)
2. Kurzowa, Z.: Illustrated Basic Dictionary of the Polish Language. Universitas, Krakow (2005)
3. Pietrzak, W.: Signed Language in School, vol. 2. WSiP, Warszawa (1992)
4. Polanski, K.: Dictionary of Syntactic Generative Verbs of the Polish Language. Ossolineum, Wroclaw-Warszawa-Krakow-Gdansk (1980-1992)
5. Pralat-Pyrzewicz, J.: Signed Language in School, vol. 3. WSiP, Warszawa (1994)
6. Szczepankowski, B.: Signed Language in School, 3rd edn., vol. 1. WSiP, Warszawa (1988)
7. Szczepankowski, B.: Sign Language. First Medical Aid. Centre of Medical Education, Warszawa (1996)
8. Szczepankowski, B.: Introductory Course of Sign Language. Warszawa (1986)
9. Szczepankowski, B.: Basics of Sign Langauge. WSiP, Warszawa (1988)

Prediction of Software Quality Based on Variables from the Development Process

Hércules Antonio do Prado[1,2], Fábio Bianchi Campos[1,3], Edilson Ferneda[1], Nildo Nunes Cornelio[4,5], and Aluizio Haendchen Filho[6]

[1] Graduate Program on Knowledge and IT Management, Catholic University of Brasilia,
SGAN 916 Av. W5, 70.790-160 – Brasília, DF, Brazil
[2] Embrapa – Management and Strategy Secretariat,
Parque Estação Biológica – PqEB s/n°, 70.770-90 – Brasília, DF, Brazil
[3] Data Processing Center of the Brazilian Senate,
Av. N2, Anexo C do Senado Federal, 70165-900 Brasília - DF
[4] Federal Bureau of Data Processing,
SGAN Quadra 601 Módulo V, 70836-900 – Brasília, DF, Brazil
[5] Centro Universitário do Distrito Federal – UDF
SEP/SUL EQ 704/904 – Conj A, 70390-045 – Brasília, DF, Brazil
[6] UNIDAVI – Universidade para o Desenvolvimento do Alto Vale do Itajaí
Rua Dr. Guilherme Guemballa - Rio do Sul - Santa Catarina, Brasil
hercules@ucb.br, fbianchi@senado.gov.br, eferneda@pos.ucb.br,
nildo.cornelio@serpro.gov.br, aluizioh@terra.com.br

Abstract. Since the arising of software engineering many efforts have been devoted to improve the software development process. More recently, software quality has received attention from researchers due to the importance that software has gained in supporting all levels of the organizations. New methods, techniques, and tools were created to increase the quality and productivity of the software development process. Approaches based on the practitioners' experience, for example, or on the analysis of the data generated during the development process, have been adopted. This paper follows the second path by applying data mining procedures to figure out variables from the development process that most affect the software quality. The premise is that the quality of decision making in management of software projects is closely related to information gathered during the development process. A case study is presented in which some regression models were built to explore this idea during the phases of testing, approval, and production. The results can be applied, mainly, to help the development managers in focusing those variables to improve the quality of the software as a final product.

Keywords: Software quality, Data mining, Regression models.

1   Introduction

The adoption of mature methodologies (e.g., RUP, SCRUM, XP) for the software development process has enabled an accumulation of information from this process

M. Graña et al. (Eds.): KES 2012, LNAI 7828, pp. 71–77, 2013.
© Springer-Verlag Berlin Heidelberg 2013

that was never seen before. It can be easily perceived that this information, adequately analyzed, can provide relevant relations to the development managers to improve the development process and the final product. Such information set can hide patterns that, if explained, can offer important input for the Software Engineering (SE) management process. Budget of projects, time schedule, and resource allocation are examples of tasks that can take advantage from these patterns.

The constant search for product quality improvement is a challenge for any development company that wishes to keep competitive in a global market characterized by the offer of a myriad of options. According to Meyer [5], there are two relevant dimensions in the software development process: the external dimension that refers to user issues, and the internal one, that relates to the developer context. These dimensions enable particular metrics for evaluating the product quality that are further discussed.

This paper presents the results of a research based in the software development experience of a Brazilian public software house, henceforth called SFPD. The focus was the relations existing among variables from the development or evolution processes and the product quality in the phases of test, homologation and production.

2   Software Quality Issues

Software quality is defined by Bartié [1] as a systematic approach focused on the achievement of software products free of flaws by preventing or eliminating defects in software artifacts during the software development process. Software quality metrics, related to the internal or the external dimensions, can be defined to help in the generation of high quality software products. For the internal dimensions, metrics like portability, reuse, and interoperability, maintainability, testability, and flexibility can be cited. For the external dimension, it could be used metrics like correction, reliability, usability, integrity, and efficiency. Some of these metrics approach aspects of economic quality. For example, COQUALMO [2] is a model for software quality that emphasizes the importance of issues like costs, timetable, and risks in development projects.

According to Yourdon [10], the main problems in software development are related to productivity, reliability, and maintainability. Such problems influence the product quality directly. Productivity is related to the availability of the resources required to meet the expectations of an increasing demanding market. Reliability is related to the level of maintenance or evolution required and is affected by errors and failure rates.

Fenton and Pfleeger [3] argue that metrics are fundamental for evaluating the performance and progress of a product development. They must assure quality and transparence to the process. Metrics guarantee that software production can be monitored, enabling a constant evaluation process, beyond the possibility of adjustments when a problem arises. According to Fenton and Pfleeger [3], defect rates in the test phase are an interesting alternative to measure the quality of a software product.

Gomes, Oliveira and Rocha [4] suggest specific metrics to the software development process for quality control. Some of these metrics are: *(i) system size in lines of code* (loc), *(ii) total project effort*, and *(iii) team experience in the programming language*. As noted by Sommerville [7], it is known that development technology, quality process, quality of the involved team, cost, time, and effort expended are parameters that influence of software product quality during the testing approval and production.

Based on these studies, the quality control team of SFPD has defined a set of variables, recognized as those that most impact the product quality (Table 1). By acting preventively on the activities related to these variables, the company drives its strategies to keep competitive in the software market.

Table 1. Independent variables. (FP: Function Points)

Variable	Type	Description
Adherence project/process	Numeric	Mean($100 - \Sigma$ Occurrences and deviations)
Type of project	Text	Software development or evolution
Systems affected	Numeric	Amount of systems depending on the project
Productivity	Numeric	Developer productivity (hours required / FP)
Team training	Text	Level of team training (High, Medium or Low)
Programming language	Text	Programming language name
Database	Text	Database name
Life cycle	Text	Life cycle adopted
Development method	Text	Software Engineering approach
Requirements volatility	Numeric	Percentage of requirements changed
Task size	Numeric	Size of the new system or the evolution (FP)
Effort	Numeric	Time spent in test (hours / *task size*)

3   Methodology

Considering the numeric nature of the dependent variables, the study was focused on finding regression relations between them and the independent variables. Analysis was carried out based on the data from the SE process adopted by SFPD. The premise for the study was that the intrinsic quality of a software product can be measured by the density of defects in the phases of test, homologation, and production. Three dependent variables related to each phase were chosen: *(i)* density of defects in the test phase, *(ii)* density of defects in the homologation phase, and *(iii)* amount of corrective maintenance requests (in Function Points) in the production phase.

The models were built by applying the algorithms provided by WEKA environment [8]. The original data set, with 183 instances, is composed by 150 instances related to development, 20 related to maintenance or evolution, and 13 non-identified.

As an exploratory study, some regression algorithms were tested in order to figure out the one that best fits in our data set, i.e., the one that gives the best correlation between dependent and independent variables. As a matter of fact, according the NFL Theorem [9], no algorithm beats all the others in all data sets. Therefore, it is important to try some algorithms in order to find the best for the used data base.

The following regression algorithms implemented in Weka were tested: Linear Regression, Least Median Squared Linear Regression (LeastMedSq), Multilayer Perceptron (MLP), Radial Basis Function (RBF), and Support Vector Regression (SMOReg). A 3-fold cross-validation approach was adopted to evaluate the quality of the models.

Since it was not known how informative the variables are w.r.t. to each type of project (development or maintenance/evolution), it were tried the following successive approximations: *(i)* the complete data set with 170 recordings, *(ii)* the data set separated by type of project, and *(iii)* the data set separated by type of project but only considering relevant independent variables as identified in the previous step (namely, *programming language, productivity, database, adherence project/process*).

4   Building the Regression Models

The models generated in each phase (test, homologation, and production), for the three approximations mentioned in the methodology (complete data set, separated data set, and focusing only in the more relevant variables) were evaluated according to the correlation coefficient. We chose this error measure instead of mean squared error, relative squared error, or any other, because correlation is scale independent and allows analysis in the well-defined interval [-1,1]. According to Santos [6], the correlation coefficient cc can be categorized as highly correlated ($|cc| > 0.7$), moderately correlated ($0.3 \leq |cc| \leq 0.7$) and weakly correlated ($|cc| < 0.3$). In this paper, we are only considering correlation coefficients greater than 0.3.

Table 2. Ranking for the complete dataset

Phase	Applied technique	Correlation coefficient	Ranking
Test	LeastMedSq	0,05	4°
	Linear regression	0,08	2°
	MLP	0,07	3°
	RBFNetwork	-0,12	5°
	SMOReg	**0,13**	1°
Homologation	LeastMedSq	**0.04**	1°
	Linear regression	0.03	2°
	MLP	-0,08	4°
	RBFNetwork	-0.16	5°
	SMOReg	0,02	3°
Production	LeastMedSq	0.17	2°
	Linear regression	0.09	4°
	MLP	0,13	3°
	RBFNetwork	0.01	5°
	SMOReg	**0,20**	1°

The results presented in the approximation with the complete dataset are shown in Table 2, where the higher correlation values are highlighted. It can be observed that positive results were not found, perhaps due to the mixing of project types in the dataset. Possibly, this aspect had an effect in the correlations between expected and found results in terms of defect density. Therefore we performed the second approximation, i.e., with the data separated by project type. Table 3 shows the results for this approximation.

It is noticeable that the amount of models with positive correlations almost doubled. However, only one model has surpassed the threshold of 0.3 established as a quality level for accepting the model. The reports from the previous analysis have shown *programming language, productivity, database,* and *adherence project/process* as the more relevant variables. Therefore, we took these variables and proceeded to the third approximation. The corresponding results are shown in Table 4.

Table 3. Ranking for the dataset separated by project types and phases

	Phase	Applied technique	Correlation coefficient	Ranking
Development	*Test*	LeastMedSq	0,00	4°
		Linear regression	0,20	2°
		MLP	0,13	3°
		RBFNetwork	-0,07	-
		SMOReg	**0,26**	1°
	Homologation	LeastMedSq	0,00	4°
		Linear regression	0,04	2°
		MLP	**0,06**	1°
		RBFNetwork	0,02	3°
		SMOReg	-0,02	-
	production	LeastMedSq	0,31	2°
		Linear regression	0,16	3°
		MLP	**0,42**	1°
		RBFNetwork	-0,07	-
		SMOReg	0,31	2°
Maintenance	*Test*	LeastMedSq	**0,20**	1°
		Linear regression	-0,14	-
		MLP	-0,05	-
		RBFNetwork	0,04	2°
		SMOReg	-0,07	-
	homologation	LeastMedSq	-0,16	-
		Linear regression	-0,16	-
		MLP	-0,06	-
		RBFNetwork	-0,34	-
		SMOReg	-0,14	-
	production	LeastMedSq	**0,08**	1°
		Linear regression	-0,07	-
		MLP	-0,06	-
		RBFNetwork	-0,13	-
		SMOReg	-0,07	-

The last approximation has shown an improvement in the production phase of development projects represented by a correlation coefficient greater than 0.7, the threshold for a highly correlated model. This result was found with the MLP model that can be used as a management tool for estimating the defects density for development projects in the production phase. By analyzing the variables weights listed as an output from Weka, it can be concluded that *programming language, productivity, database,* and *adherence project/process* are variables that can affect the software quality in the production phase of a development process. The lack of results for other types of systems and phases can be a consequence of the scarcity of data for this kind of analysis.

Table 4. Ranking for the dataset separated by project types and phases considering relevant independent variables

	Phase	Applied technique	Correlation coefficient	Ranking
Development	*Test*	LeastMedSq	-0,18	-
		Linear regression	**0,20**	1°
		MLP	0,16	3°
		RBFNetwork	-0,08	-
		SMOReg	**0,20**	1°
	homologation	LeastMedSq	0,00	-
		Linear regression	0,13	2°
		MLP	**0,14**	1°
		RBFNetwork	0,09	3°
		SMOReg	0,08	4°
	production	LeastMedSq	0,34	2°
		Linear regression	0,21	4°
		MLP	**0,71**	1°
		RBFNetwork	0,00	-
		SMOReg	0,32	3°
Maintenance	*Test*	LeastMedSq	0,15	2°
		Linear regression	-0,25	-
		MLP	-0,01	-
		RBFNetwork	**0,17**	1°
		SMOReg	-0,10	-
	homologation	LeastMedSq	0,00	-
		Linear regression	-0,05	-
		MLP	-0,04	-
		RBFNetwork	-0,43	-
		SMOReg	-0,18	-
	production	LeastMedSq	0,00	-
		Linear regression	-0,03	-
		MLP	-0,10	-
		RBFNetwork	-0,10	-
		SMOReg	-0,20	-

5   Conclusion and Further Works

Establishing relations among variables from a previous to a particular phase can provide useful feedback for a software engineering manager. On the basis of observations taken from the homologation, test, and development phases, and knowing the quality of a particular software in the production phase, the manager could define focused interventions to assure better quality in team results. The findings of this study have shown that this approach can lead to interesting results. However, it must be noticed that some techniques corroborate this practice in Software Engineering while other do not. This possibly happened due to scarcity of data available for analysis. In this sense, this work shall be considered as an exploratory study aiming at designing an analysis approach in order to predict the software quality in each phase of its life cycle. We found that in order to achieve meaningful results, a larger dataset is recommended.

In the present work, only the intrinsic software quality was studied, that is, the software quality inside the development environment. It is important to further expand this study in order to consider the user's perception of software quality.

References

1. Bartié, A.: Garantia da qualidade de software: adquirindo maturidade organizacional. Elsevier, Rio de Janeiro (2002)
2. Chulani, S., Steece, B., Boehm, B.: Determining software quality using COQUALMO. In: Blischke, W., Murthy, D. (eds.) Case Studies in Reliability and Maintenance. Wiley, Sidney (2002)
3. Fenton, N., Pfleeger, S.L.: Software Metrics - A Rigorous & Practical Approach. PWS Publishing Company, Boston (1997)
4. Gomes, A., Oliveira, K., Rocha, A.R.: Avaliação de Processos de Software Baseada em Medições. In: Proceedings of the XV Brazilian Symposium on Software Engineering, Rio de Janeiro, pp. 84–99 (2001),
http://www.lbd.dcc.ufmg.br:8080/colecoes/sbes/2001/006.pdf
5. Meyer, B.: Object Oriented Software Construction. Prentice-Hall, New Jersey (1988)
6. Santos, C.: Estatística Descritiva: Manual de auto-aprendizagem. Edições Sílabo, Lisbon (2007)
7. Sommerville, I.: Software Engineering, 8th edn. Addison-Wesley (2006)
8. Witten, I.H., Frank, E.: Data Mining: Practical machine learning tools and techniques, 2nd edn. Morgan Kaufmann, San Francisco (2005)
9. Wolpert, D.H.: The lack of a priori distinctions between learning algorithms. Neural Computation 8(7), 1341–1390 (1996)
10. Yourdon, E.: Modern Structured Analysis. Prentice Hall (1988)

Mining High Performance Managers
Based on the Results of Psychological Tests

Edilson Ferneda[1], Hércules Antonio do Prado[1,2],
Alexandre G. Cancian Sobrinho[1], and Remis Balaniuk[1]

[1] Graduate Program in Knowledge Management and IT, Catholic University of Brasilia
SGAN 916 - Módulo B - Sala A-111, 70.790-160, Brasília, DF, Brazil
[2] Embrapa - Management and Strategy Secretariat,
Parque Estação Biológica - PqEB s/n, 70.770-901, Brasília, DF, Brazil
eferneda@pos.ucb.br, hercules@ucb.br,
alexandre.gsc@hotmail.com, remis@ucb.br

Abstract. Selecting high performance managers represents a risky task mainly due to the costs involved in a wrong choice. This fact led to the development of many approaches to select the candidates that best fit into the requirements of a certain position. However, defining what are the most important features that condition a good personnel performance is still a problem. In this paper, we discuss an approach, based on data mining techniques, to help managers in this process. We built a classifier, based in the Combinatorial Neural Model (CNM), taking as dependent variable the performance of managers as observed along their careers. As independent variables, we considered the results of well-known psychological tests (MBTI and DISC). The rules generated by CNM enabled the arising of interesting relations between the psychological profile of managers in their start point in the company and the quality of their work after some years in action. These rules are expected to support the improvement of the selection process by driving the choice of candidates to those with a best prospective. Also, the adequate allocation of people – the right professional in the right place - shall be improved.

Keywords: Knowledge-Based Systems, Combinatorial Neural Model, Psychological Evaluation, Performance Management.

1   Introduction

The common practice of managers' selection is based on the evaluation of personal attributes of the candidates. Quantitative and qualitative methods are applied in order to figure out how the candidates fit in the expected profile. The methods and their metrics vary according to the characteristics of the function and the strategic importance of the position to be filled. It is well known that the selection costs may vary according to the function to be performed and that the damage caused by inappropriate selection is directly proportional to the hierarchical level of the position. So, constantly, companies seek for alternative tools and techniques that can improve the selection process.

M. Graña et al. (Eds.): KES 2012, LNAI 7828, pp. 78–87, 2013.

This paper examines the relations between the results of two psychological tests and the observed performance of a set of managers in order to support the personnel managers in improving both the people selection as the results of their career. A case study was carried out in a large company that adopts MBTI [1] and DISC [4] tests to their candidates and proceeds to a systematic manager's evaluation. In the case studied, the recruitment/selection and staff development generate results that require a subjective evaluation of human resources of the company.

The most important challenge for the personnel management is to build an accurate model able to identify the relationship between psychological profiles and staff performances. The main purpose of this model is to allow a less subjective analysis of the results of the tests applied during the recruitment and selection process.

2   Psychological Tests Used

2.1   MBTI

The application of psychological analysis to seek professionals that best fit the labor demands were intensified due to the economic difficulties brought by the Second World War. In 1943 Katherine Briggs and Isabel Myers proposed a set of questions that could help in adjusting people without previous experience to a particular activity that later became known as MBTI (Myers-Briggs Type Indicator). Currently, MBTI is widespread and used in corporate environment to define teams and improve its management [1]. The concept of psychological types in the MBTI is related to the idea that individuals feel more comfortable with certain ways of thinking and acting [2]. The method is based on four pairs of exclusive preferences or "dichotomies" with regard to specific situations: Extroversion (E) / Introversion (I), Sensory (S) / Intuition (N), Thinking (T) / Feeling (F), and Judgment (J) / Perception (P).

Note that the terms for each dichotomy have specific technical meanings related to the MBTI, which differ from their everyday meaning. For example, people with a preference for judgment over perception are not necessarily more critics or less perceptive. It just means that they use more naturally judgment than perception. The same person can use the perception, if necessary, but with more effort. MBTI does not measure the attitudes; it just shows which preferences stand out over others. As the dichotomies are based on scales of varying intensity, a naturally extroverted person may be considered introverted when compared to someone else who has a degree of extroversion bigger than him/her. Sixteen psychological types (eight sensory and eight intuitive) can arise by combining the preferences for each situation (Table 1). Also, these types can be categorized as traditionalists, visionaries, catalysts, and tactical, depending on the predominance of determinate dichotomies.

2.2   DISC

In the early 20's, the psychologist William Marston sought to explain the emotional responses of people. His concepts were originally intended to understand and systematize models of individuals' interaction with their environment, identifying the different mechanisms used by people in pursuit of pleasure and harmony [4].

Table 1. Two possible categorizations of psychological types [3]
(1 = Traditionalist, 2 = Visionary, 3 = Catalysts, 4 = Tactical)

	... with thinking	... with feeling
Sensory	ISTJ: Introverted with judgment[1] ISTP: Introverted with perception[4] ESTP: Extroverted with perception[4] ESTJ: Extroverted with judgment[1]	ISFJ: Introverted with judgment[1] ISFP: Introverted with perception[4] ESFP: Extroverted with perception[4] ESFJ: Extroverted with judgment[1]
	... with feeling	**... with thinking**
Intuitive	INFJ: Introverted with judgment[3] INFP: Introverted with perception[3] ENFP: Extroverted with perception[3] ENFJ: Extroverted with judgment[3]	INTJ: Introverted with judgment[2] INTP: Introverted with perception[2] ENTP: Extroverted with perception[2] ENTJ: Extroverted with judgment[2]

Marston developed a technique related to stimulus and response to measure the types of behavior that he was trying to describe. He chose the following factors to measure: Dominance, Influence, Steadiness, and Conformity. DISC is an acronym of these factors and represents a four quadrant behavioral model that maps the behavior of individuals in their environment or in a specific situation. DISC takes into account styles and behavioral preferences. DISC has become, probably, the most adopted theory to assess the psychological profile in the world. In the DISC terminology, the behavior is defined as the sum of the different types of responses of a person to various stimuli. It is important to point out that there is not a factor better or worse than other, a DISC profile only reports a behavioral style, and each style has its inherent strengths as well as their weaknesses.

It is common to find types of behaviors represented by only one factor, for example, with a high "S". However, the most common profiles have a combination of these factors, leading to a more complex interpretation as the various factors are combined. The factors considered have the following meanings:

- People with high "D" (*Dominance*) are very active in dealing with problems and challenges, while those with low "D" are people who prefer to do more research before taking a decision;
- People with high levels of "I" (*Influence*) influence others with words and actions and tend to be emotional; those with low values of "I" influence more through data and facts rather than feelings;
- People with high values in "S" (*Steadiness*) want a steady environment, security, and do not like sudden change; on the other hand, low values in "S" describe people that like change and variety; and
- People with a high value on "C" (*Compliance*) adhere to the rules, regulations and structure; those with low values in "C" challenge the rules, want independence and are described as independent and careless with details.

The real power of DISC comes from its ability to interpret the relationship between these factors. For example, if a person has a predominant Dominance and a similarly high level of Influence, he/she will behave quite differently in relation to an individual equally dominant, but that does not have high influence. The factors can be combined to provide (theoretically) around one million different profiles.

3   Methodological Issues

The Combinatorial Neural Model (CNM) is a hybrid model for intelligent systems that adopt the better of two worlds: the learning ability from neural networks and the expressivity of symbolic knowledge representation. It was proposed by Machado and Rocha [5] as an alternative to overcome the black box limitation of the Multilayer Perceptron [6], by applying the neural network structure along with a symbolic processing. CNM can identify regularities among input vectors and output values, performing a symbolic mapping. It uses supervised learning and a feedforward topology with three layers (see Figure 1):

- The input layer, in which each node corresponds to a triple object-attribute-value that describes some dimension (here called *evidence* and denoted by e) in the domain;
- An intermediary or combinatorial layer with neurons connected to one or more neurons in the input layer, representing the logical conjunction AND; and
- The output layer, with one neuron for each possible class (here called *hypothesis* and denoted by h), that is connected to one or more neurons in the combinatorial layer by the logical disjunction OR.

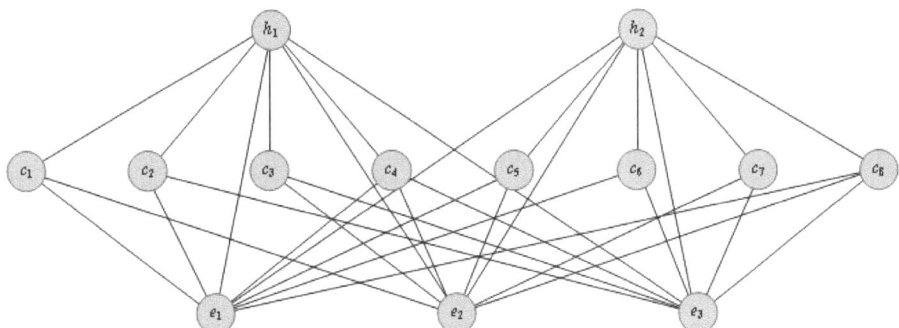

Fig. 1. An example of Combinatorial Neural Model

The synapses may be inhibitory or excitatory and have assigned a weight between zero (uncommitted) and one (fully connected). The network is created as follows:

(i)   one neuron in the input layer for each evidence in the training set;
(ii)   a neuron in the output layer for each class in the training set; and
(iii)   one neuron in the combinatorial layer for each possible combination of evidences (with order ≥ 2) from the input layer.

Combinations of order = 1 are connected directly to the output layer. CNM is trained according to the learning algorithm shown in Figure 2.

PUNISHMENT_AND_REWARD_LEARNING_RULE
(i)   **Set** the initial value of the accumulator in each arc to zero;
(ii)   **For** each example from the training set, **do**:
Propagate the evidences from input nodes to the output layer;
(iii)   **For** each arc arriving to a neuron in the output layer, **do**:
If the output neuron corresponds to a correct class
Then backpropagate from this neuron to the input nodes increasing the
accumulator of each traversed arc by its evidential flow (reward)
Else   backpropagate decreasing the accumulators (punishment).

Fig. 2. Learning algorithm for CNM

For the sake of simplicity, without quality losses, we used a constant value 1 for the evidential flow. For the CNM training, the examples are presented to the input layer, triggering a signal from each neuron matched to the combinatorial layer, having their weights increased. Otherwise, their weights are weakened. After the training, the accumulators associated to each arc in the output layer will belong to the interval $[-c, c]$, where c is the number of cases in the training set. After the training process, the weights network is pruned as follows:

(i)   Remove all arcs arriving to the output layer with accumulators below a threshold specified by the user and

(ii) Remove all neurons and arcs from the input and combinatorial layers disconnected after the first step.

The relations that remained after the pruning are considered rules in the application domain. Two basic learning metrics are applied during the model generation that allow the evaluation of the resulting rules:

- *Rule confidence (CI)* is the percentage that expresses the number E of examples in which the rule R is valid with respect to the total cases of class C.
- *Rule support* is the percentage that expresses the number of examples E in which the rule R holds with respect to the total amount of cases (T).

This study assumes that there is a relationship between behavioral and psychological profiles and the performance in the job. Also, that this relationship varies according to the profile of the group in which the professional is inserted. The data analyzed includes

- 85 MBTI assessments, with their degrees of extroversion, introversion, sensory, intuition, thinking, feeling, judgment, and identified perception,
- 60 DISC assessments, covering the degree of dominance, influence, steadiness and compliance of internal and external profiles, and
- Data of the results of competency and performance evaluation related of the years 2003 to 2008.

4   Case Study

4.1   Data Exploration

The data concerning to the MBTI assessments of employees were summarized and grouped according to their behavioral characteristics (see Table 2).

Table 2. Types distribution

Type	Mood	Company (%)	Description
SJ	Traditionalist	61.18	Stabilizer, consolidating
NT	Visionary	31.76	System Architect, builder
NF	Catalyst	5.88	Idealist, communicative, motivating
SP	Tactical	1.18	Solve problems, dealer, reliever

Notice that 61.18% of the evaluated staff has SJ (Traditionalists) as predominant temper. Employees with the characteristic NT (visionaries) totalize 31.76%. The types NF (catalysts) totalize 5.88% and SP (tactical) only 1.18%. Thus the managers are predominantly traditionalist and visionary. These two temperaments totalize 92.94% of the total, while the tactical and catalyst temperaments totalize 7.06%. So, according the predominant temperament, the management is *(i)* very focused on meeting deadlines, solving problems in stable and structured environments, *(ii)* focused in details and planning, seeking recognition and respect, and *(iii)* make decisions based on realities and facts, striking features traditionalists (SJ). It was also observed a high incidence of visionary employees with (NT) that search for respect by means of innovations that are implemented and autonomy. In both temperaments, logic and rationality are dominant.

The data for the DISC assessments were grouped also according to their behavioral characteristics. Table 3 summarizes the DISC distributions of analyzed staff. The amount of employees having performance evaluation, during the period considered (2003 to 2008), was 545, 558, 537, 555, 498, and 658, respectively.

Table 3. Grouping the results of DISC assessments

Classification	Internal class profile (%)				External class profile (%)				Summary class profile (%)			
	D	I	S	C	D	I	S	C	D	I	S	C
Very high	12,3	19,3	7,0	3,5	1,8	10,5	0,0	7,0	1,7	11,7	1,7	6,7
High	7,0	24,6	28,1	24,6	29,8	56,1	7,0	38,6	18,3	48,3	13,3	35,0
Average	56,1	35,1	52,6	49,1	24,6	12,3	43,9	35,1	50,0	21,7	51,7	40,0
Low	24,6	17,5	12,3	22,8	40,4	17,5	43,9	17,5	30,0	16,7	31,7	18,3
Very Low	0,0	3,5	0,0	0,0	3,5	3,5	5,3	1,8	0,0	1,7	1,7	0,0
TOTAL	100,0	100,0	100,0	100,0	100,0	100,0	100,0	100,0	100,0	100,0	100,0	100,0

4.2   Data Preparation

The results of MBTI and DISC assessments applied to employees with senior management level were selected. The results from the evaluations (Skills and Goals) carried out during 2003 to 2008 were also selected. The information generated by MBTI, DISC, and performance evaluations has been consolidated in the training set, keeping only 45 employees with performance assessment MBTI and DISC simultaneously.

The inputs were discretized as follows:

- For the dichotomies intensities of MBTI, it was adopted the values *Mild* for the interval [50%, 59%), *Moderate* for [59%, 75%), *Clear* for [75%, 90%) and *Very Clear* for percentages and> 90%;
- For DISC assessment, it was adopted the values *Very High* for percentages ≥ 80%, *High* for the interval [60%, 80%), *Medium* for [40%, 60%), *Low* for [20%, 40%), and *Very Low* for percentages <20%; and
- For competence and performance it were adopted *Lousy* for percentages <50%, *Low* for [50%, 60%), *Regular* for [60%, 70%), *Good* for [70%, 80%), *Very Good* for [80%, 90%), and *Excellent* for ≥ 90%.

The entire training dataset was inspected for quality issues and the results showed in Table 4.

Table 4. Data quality report

Tables	Data Quality	Contingency
Staff	The data are quite consistent and cover all active and retired employees of the company	There was no need for additional treatment as the data has excellent quality
	Some employees have repeated their registration in the database because they got off and returned the company	Registration details of the employees began to be controlled by a unique key
MBTI Tests	Only employees at the managerial level or graduate performed tests	There were used for analysis only the employees who had performed the tests.
	Some of the managerial level employees and managers had not conducted the test	Tests were performed on all employees who had not performed the evaluations in order to complete the database
Tests DISC	Some of the managerial level employees and managers had not conducted the test	The data was discarded
Individual Performance	The data are quite consistent and cover all active and retired employees	No further data treatment was necessary , as they have excellent quality
	The 2008 data were incomplete	We requested the release of 2008 data in order to include them in the model

4.3   Modeling

As previously mentioned, the modeling phase was developed by applying CNM. First, a unique classifier was built taking all variables from MBTI and DISC as input variables in just one training dataset. However, it was not reached a good level of quality what led to the separation of the training dataset into two separate files: one for each assessment tool. The most significant results obtained by applying CNM (separately, for MBTI and for DISC) are shown in Table 5.

These results correspond to a descriptive analysis of the relationships between the variables of psychological tests and performance evaluations. In order to define a "cut line" it was arbitrarily defined the confidence higher than 60% as expressing the most interesting relations that should be analyzed. The MBTI model shows a degree of confidence that varies between 64.64% and 77.78%. For the DISC model, the confidence level ranges from 60% to 83.33%.

Table 5. Results of modeling with Confidence ≥ 60%

	#	If ...	then PERF = ...	Class	Confidence (CI)	Cases	Support
MBTI	1	Perception = N – Mild	Very Good	46,05%	77,78%	7	20,00%
	2	Decision = T - Moderate	Very Good	46,05%	76,92%	10	28,57%
	3	Decision = T - Very Clear	Very Good	46,05%	63,64%	7	20,00%
	4	Perception = N - Moderate	Very Good	46,05%	63,64%	7	20,00%
	5	Perception = S - Moderate & Decision = T – Clear	Excellent	35,53%	63,64%	7	25,93%
	6	Energy = E – Mild	Excellent	35,53%	63,64%	7	25,93%
DISC	1	DISC-I-I-Class = High	Very Good	46.05%	83,33%	10	28.57%
	2	DISC-D-I-Class = Low & DISC-D-R-Class = Low	Very Good	46.05%	80.00%	8	22.86%
	3	DISC-C-I-Class = Average & DISC-D-R- Class = Average & DISC-C-R- Class = Average	Very Good	46.05%	77.78%	7	20.00%
	4	DISC-D-E- Class = Average & DISC-S-R- Class = Average	Very Good	46.05%	77.78%	7	20.00%
	5	DISC-I-I- Class = Average & DISC-C-R- Class = Average	Very Good	46.05%	77.78%	7	20.00%
	6	DISC-D-I- Class = Low & DISC-D-E- Class = Low	Very Good	46.05%	77.78%	7	20.00%
	7	DISC-D-R- Class = Average & DISC-C-R- Class = Average	Very Good	46.05%	76.92%	10	28.57%
	8	DISC-C-E-Class = Average & DISC-I-R-Class = High	Very Good	46.05%	76.92%	10	28.57%
	8	DISC-C-E-Class = Average & DISC-I-R-Class = High	Very Good	46.05%	76.92%	10	28.57%

4.4   Evaluation

In this section the classification results (Table 5) are discussed in order to evaluate the representativeness of the model. The results were promising and positive, since in both cases were found rules with the degrees of confidence above 60%.

Results for MBTI

The modeling with MBTI data generated a model more concise than the one from DISC. Two rules stand out from the results by reaching the higher confidence levels. They points out that employees characterized by *(i)* a mild form of perception and *(ii)* a moderate decision process tend to have a very good performance, the former with $CI = 77.78\%$ and the latter with $CI = 76.92\%$. People in the first case present a good balance between intuition and sensation. They seek the meanings, relationships, and future possibilities of the information received. They tend to perceive the phenomena (people, events, and objects) subjectively. However, by having a light characteristic, they tend to have a good balance in searching for data and facts due to the use of perception. In the second case, the people discriminate, judge, and classify the phenomena from the logic of reason, seeking to evaluate objectively the 'pros' and 'cons' of the nature of these phenomena. For the remaining rules, it was observed that good results in performance are characterized by *(i)* decision process moderate or very clear, *(ii)* perception moderate, *(iii)* perception moderate and decision clear, and *(iv)* energy mild. These rules characterize two psychological types, visionary and traditionalist. The first refers to professionals who seek new possibilities, not losing focus on the reality. The use of reason and logic is also a major factor in these professionals. The second are those that are not too open to changes, keeping the focus in doing the ordinary in the best possible way.

Results for DISC

Although the baseline of $CI = 60\%$ adopted for characterizing interesting rules, the results for DISC started in $CI = 76.92\%$. The model highlights two rules with confidence equal to or greater than 80 and suggests the following relationship:

- Employees characterized by a high internal influence tend to have a good performance result with $CI = 83.33\%$; people with this feature are enthusiastic with a great ability to motivate and influence people in the search for results through cooperation;
- Employees characterized by a low level of internal dominance tend to have a very good performance result with $CI = 80.00\%$; people with this characteristic are, in general, conservative, slightly hesitant, reserved, moderate, non-aggressive, diplomats, and careful; and
- Combining high values of influence with low values of dominance, we have people with potentially large capacity of motivation and leadership coupled with a tendency to diplomacy and low friction - these are important characteristics for leaders who want to form teams with good synergy and promote a collaborative teamwork.

Discussion

In general, the results adhere to the commonsense in personnel management. It may represent that CNM model effectively represents the relations between variables from the assessment tools and the performance results. However, the higher importance of influence over dominance, as shown in rule 1 of DISC, does not match with the expected behavior.

Based on the results obtained it can be concluded that this approach has potential to produce excellent results in the recruitment process, selection, training and development of the company under study. Thus, it seems promising to extend this technique to other operating levels of the company. The results were presented to the company's board in order to demonstrate the potential of the adopted approach and received a positive evaluation, pointing out to deepen the assessment process.

5   Conclusions and Further Works

A consistent relation between the results of MBTI and DISC were observed, showing that the adopted approach could be considered a promising research path. In this sense, more robust results would require the extension of this study to a wider data universe, involving other companies. So, this discussion can only be considered with regard to the case study presented.

The adopted model generated rules with a reasonable degree of accuracy, despite the reduced amount of data. The exploratory nature of the study does not allow terminative conclusions, but provides evidence that its extension to a confirmatory study can generalize the relations found. Thus, it is understood that future works should include:

- The design and realization of a confirmation experiment of the found relations, with an adequate sample characterization,
- To propose subsidies for changing policies and practices of performance management personnel,
- The application of the used method with data generated by other tools of psychological and behavioral assessment, such as, the EQI (Coefficient of Emotional Intelligence), and
- The construction of decision support systems in the organization development area.

References

1. Myers, I.B., McCaulley, M.H., Quenk, N.L., Hammer, A.L.: MBTI Manual (A guide to the development and use of the Myers Briggs type indicator), 3rd edn. Consulting Psychologists Press (1998)
2. Keirsey, D., Bates, M.: Please Understand Me: Character and Temperament Types. Prometheus Nemesis, Del Mar (1978)
3. Myers, I.B., Myers, P.B.: Gifts differing. Consulting Psychologists Press, Palo Alto (1997)
4. Matos, J., Portela, V.: Talent for Life – What to do for discovering and enhancing your talents and have a productive and delightful life. HLCA - Human Learning, Rio de Janeiro (2001) (in Portuguese)
5. Machado, R.J., Rocha, A.F.: Handling Knowledge in High Order Neural Networks: The Combinatorial Neural Model. Technical Report. CCR-076. IBM - Centro Tecnológico Rio, Rio de Janeiro (1989)
6. Haykin, S.: Neural Networks: A Comprehensive Foundation. Prentice Hall (1999)

Semantics Preservation in Schema Mappings within Data Exchange Systems

Tadeusz Pankowski

Institute of Control and Information Engineering,
Poznań University of Technology, Poland
tadeusz.pankowski@put.poznan.pl

Abstract. We discuss the problem of preservation of semantics of data
in data exchange systems. Semantics of data is defined in a source database
schema by means of integrity constraints and we expect that the semantics will be preserved after transforming the data into a target database.
The transformation is defined by means of schema mappings. In order to
reason about soundness and completeness of such mappings with respect
to semantics preservation, we propose a method of developing a knowledge base capturing both databases and mappings in a data exchange
system.Then formal reasoning about consistency of schema mappings
with integrity constraints of databases and with knowledge of application domain can be performed.

1 Introduction

Data exchange is the problem of transforming data structured under a source
schema into an instance of a target schema that reflects the source data as accu-
rately as possible. The transformation is specified by means of source-to-target
dependencies (STDs) and can be performed using the classical chase procedure
[1,10]. The data exchange problem can be also perceived as a part of complex
data integration activities both in materialized integration [4] and in virtual integration (e.g. in P2P environment) [5]. We identified this problem in our SixP2P
project for semantic integration of data in P2P environment [6,14].

One of the issues faced in data exchange is the preservation of data semantics.
Unfortunately, very often the semantics behind the schema is unclear and two
syntactically similar schemas can be semantically different. To deal with the
problem data repositories, such as relational or XML databases, are enriched
with semantic knowledge by a semantic annotation in a shared domain ontology
[3,9,18] or by translating into a formal ontology-oriented system in which the
formal reasoning is possible [7,17,12]).

In this paper we propose a method for representing a data exchange system in a formalized, knowledge based system. There are three main points in
our approach: (1) We discuss an extended knowledge base for representing relational database. The representation captures relational schema and integrity
constraints (consisting of primary- and foreign key dependencies) as well as the
database instance. We show that the semantic model of the database can be made

M. Graña et al. (Eds.): KES 2012, LNAI 7828, pp. 88–97, 2013.

using a slight extension of OWL 2 EL profile of OWL 2 ontology [13]. Axioms in the knowledge base are specified by first order rules (called ORL (*Ontology Rule Language*) rules) being an extended version of DLP [11]. There is a clear distinction in the knowledge base between axioms representing integrity constraints of databases and the domain knowledge describing interrelationships between concepts from different databases (in particular, from the source and the target databases in the data exchange system). (2) We show how STDs defining schema mappings can be translated into ORL rules. In this way a unified rule-oriented representation of both knowledge base axioms and schema mapping dependencies is achieved. This is the prerequisite for reasoning about schema mappings. (3) We discuss, on a simple example, the application of the developed method for reasoning about schema mappings in a data exchange scenario. Firstly, the STDs must be consistent with integrity constraints of the target database and with the general knowledge about application domain. Secondly, schema mappings should capture as many data as possible. In other words, we are looking for a possibly maximal part of the source database that can be mapped to the target database.

In Section 2 a problem of data exchange for relational databases is reviewed. A knowledge base for the data exchange system is discussed in Section 3. In Section 4 we study on an example a semantics preserving data exchange, focusing on its consistency and maximality. Section 5 concludes the paper.

2   Data Exchange for Relational Databases

2.1   Relational Database

A relational database schema arises usually from a conceptual model created using Entity-Relationship approach. Thus, we can assume without loss of generality that all relational schemas are in Boyce-Codd Normal Form (BCNF) [1]. In such schemas only key dependencies and foreign key dependencies (representing inclusion dependencies) are significant. Moreover, we can assume that both primary keys and foreign keys consists only of one column (attribute), otherwise a surrogate primary key can be introduced.

A *(relational) database schema* is a pair

$$DB = (\mathbf{R}, PKey \cup FKey),$$

where $\mathbf{R} = \{R_1, \ldots, R_n\}$ is a *relational schema* and $PKey \cup FKey$ is a set of *integrity constraints*: *primary key-* ($PKey$) and *foreign key-* ($FKey$) dependencies, respectively. Each *relational symbol* $R \in \mathbf{R}$ has a *sort*, which is a finite set U of *attributes*, $att(R) = U$. Sorts of relational symbols are pairwise disjoint. A *primary key dependency* is an expression $pkey(R, A)$, where $R \in \mathbf{R}$, and $A \in att(R)$ (meaning that A is a primary key for R). A *foreign key dependency* is an expression $fkey(R, A, R', A')$, where $R, R' \in \mathbf{R}$, and $A \in att(R)$, $A' \in att(R')$, (meaning that A from R references the primary key A' of R').

Let Const be a set of *constants*. A *tuple* of sort $U = \{A_1, \ldots, A_k\}$ is a function r from U to Const, written as a set of pairs $r = [A_1 : c_1, \ldots, A_k : c_k]$, where

$c_i = r.A_i \in$ Const. An *instance I of a database schema DB*, denoted DB^I, is a function that associates to each relational symbol $R \in \mathbf{R}$ a finite set of tuples of the sort $att(R)$, and such that all integrity constraints in DB are satisfied; i.e.:
(a) for each $pkey(R, A) \in PKey$, values of A uniquely identify tuples in R^I, and
(b) for each $fkey(R, A, R', A') \in FKey$, if $r \in R^I$ then there is exactly one such tuple $r' \in R'^I$ that $r.A = r'.A'$.

2.2 Data Exchange Setting

A *data exchange setting* (or *schema mapping*) is a triple

$$\mathcal{M} = (DB_s, DB_t, \Sigma_{st}),$$

where $DB_s = (\mathbf{R}_s, IC_s)$ and $DB_t = (\mathbf{R}_t, IC_t)$ are source and target database schemas, and Σ_{st} is a set of a source-to-target dependencies (STDs) over $\mathbf{R}_s$ and $\mathbf{R}_t$. The *data exchange problem* is defined as follows [10]: Let I be an instance of DB_s. Find an instance J of DB_t such that the pair (I, J) satisfies Σ_{st}, $(I, J) \models \Sigma_{st}$. Such a J, if it exists, is called a *solution* for I with respect to $\mathcal{M}$. A way for finding solutions is based on *chasing* [1], and a solution obtained in chasing is called a canonical *universal solution* [10,8].

Each STD is a formula $\forall \mathbf{u}, \mathbf{v}.(\varphi_s(\mathbf{u}, \mathbf{v}) \Rightarrow \exists \mathbf{w}.\psi_t(\mathbf{u}, \mathbf{w}))$, where $\varphi_s(\mathbf{u}, \mathbf{v})$ is a conjunction of atomic formulas over $\mathbf{R}_s$ and $\psi_t(\mathbf{u}, \mathbf{w})$ is a conjunction of atomic formulas over $\mathbf{R}_t$. Each integrity constraint is a formula of the form

$$\forall \mathbf{u}.(\varphi(\mathbf{u}) \Rightarrow u = u') \qquad - \textit{equality-generating-dependency or}$$
$$\forall \mathbf{u}, \mathbf{v}.(\varphi(\mathbf{u}, \mathbf{v}) \Rightarrow \exists \mathbf{w}.\psi(\mathbf{u}, \mathbf{w})) \quad - \textit{tuple-generating-dependency,}$$

where $\varphi(\mathbf{u})$, $\varphi(\mathbf{u}, \mathbf{v})$ and $\psi(\mathbf{u}, \mathbf{w})$ are conjunctions of atomic formulas over $\mathbf{R}_s$ (or $\mathbf{R}_t$), and u, u' are variables in $\mathbf{u}$. The former are used to specify primary key dependencies while the latter, to specify foreign key dependencies.

For example, the following two dependencies specify that the primary key for Student(SId, Name, Type) consists of the first column:

$$\forall u, v_1, v_2, w_1, w_2.(\text{Student}(u, v_1, w_1) \wedge \text{Student}(u, v_2, w_2) \Rightarrow v_1 = v_2)$$
$$\forall u, v_1, v_2, w_1, w_2.(\text{Student}(u, v_1, w_1) \wedge \text{Student}(u, v_2, w_2) \Rightarrow w_1 = w_2),$$

and the following formula specifies that the second column of the following relation Exam(EId, ESId, Course, Grade) is a foreign key referring to the primary key of Student.

$$\forall u_1, u_2, u_3, u_4.(\text{Exam}(u_1, u_2, u_3, u_4) \Rightarrow \exists u_5, u_6.\text{Student}(u_2, u_5, u_6)).$$

3 A Knowledge Base Representing Data Exchange System

In [12] an *extended knowledge base*, devoted to represent ontological knowledge described by means of description logic (DL) [2], is defined as a triple

$$\mathcal{KB} = (\mathcal{S}, \mathcal{C}, \mathcal{A}),$$

where: (a) $\mathcal{S}$ is a finite set of *standard* TBox axioms, (2) $\mathcal{C}$ is a finite set of *integrity constraint* TBox axioms, and (3) $\mathcal{A}$ is a finite set of ABox facts.

Distinguishing between *standard* ($\mathcal{S}$) and *integrity constraint* ($\mathcal{C}$) axioms depends on whether an axiom imply new facts or not. Axioms in $\mathcal{S}$ can act as *deductive rules* and can be used for generating new facts, while axioms in $\mathcal{C}$ can be used for *checking* consistency of a given set of facts. The treatment of an axiom as a standard one or as an integrity constraint depends mainly on the requirements and the assumptions of the application domain and is often not inherent to a particular axiom.

In our approach, a database DB is represented as a $\mathcal{KB}$, where $\mathcal{C}$ represents the integrity constraints of DB (then relational schema is also a kind of integrity constraints), $\mathcal{A}$ represents an instance of DB, and $\mathcal{S}$ is empty. If $\mathcal{KB}$ represents a set of databases (in a data exchange system it encompasses the source and the target databases) then $\mathcal{S}$ contains knowledge about interrelationships between concepts from these databases. In this way we achieve the soundness w.r.t. semantics preservation but not completeness (see [15]).

3.1   Representing a Relational Database by a Knowledge Base

For a relational database schema DB, and its instance I, we create a knowledge base $\mathcal{KB} = (\emptyset, \mathcal{C}, \mathcal{A})$. We assume that all integrity constraints of the DB, i.e. constraints concerning attributes, primary keys and foreign keys, are mapped to integrity constraint axioms ($\mathcal{C}$). The set of standard axioms ($\mathcal{S}$) is empty at the beginning, and will be later enriched with the application domain knowledge. The set $\mathcal{A}$ of assertions consists of facts representing the database instance.

As a formal framework for the knowledge base we assume a description logic with a concrete domain ADom.

Representation of Database Schema. Let $DB = (\mathbf{R}, PKey \cup FKey)$ be a relational database schema. Translation of this schema to a set $\mathcal{C}$ of integrity constraint axioms is given in Table 1, and the specification is: a formal notation in description logic (DL), an expression in OWL 2 language (OWL 2), and in the first order rules (ORL).

1. For each relational symbol $R \in \mathbf{R}$ there is a class (atomic concept) $C_R \in \mathsf{N}_C$.
2. For each attribute $A \in att(R)$ there is a data property (concrete role) $D_A \in \mathsf{N}_{DP}$, and:
   - domain of D_A is C_R (Table 1(1)),
   - range of D_A is ADom (Table 1(2)),
   - each individual of class C_R has at least one value of D_A (Table 1(3)),
   - each individual of class C_R can have at most one value of D_A, i.e. D_A is a functional data property (Table 1(4)).
3. If A is the primary key of R, i.e. $pkey(R, A) \in PKey$, then the data property D_A is the key of the class C_R (Table 1(5)).
4. If A is a foreign key of R referencing to A' in R', i.e. $fkey(R, A, R', A') \in FKey$, then there is an object property (abstract role) $P_A \in \mathsf{N}_{OP}$, and:

Table 1. Representation of database schema

	DL	OWL 2 EL	ORL
(1)	$\exists D_A.\mathsf{ADom} \sqsubseteq C_R$	$\mathtt{DataPropertyDomain}(D_A,\ C_R)$	$D_A(x,v) \wedge \mathsf{ADom}(v) \Rightarrow$ $C_R(x)$
(2)	$\top \sqsubseteq \forall D_A.\mathsf{ADom}$	$\mathtt{DataPropertyRange}(D_A,\ \mathsf{ADom})$	$D_A(x,v) \Rightarrow \mathsf{ADom}(v)$
(3)	$C_R \sqsubseteq \exists D_A.\mathsf{ADom}$	$\mathtt{SubclassOf}(C_R,$ $\quad\mathtt{ObjectSomeValuesFrom}($ $\quad\quad D_A, \mathsf{ADom}))$	$C_R(x) \Rightarrow \exists v.D_A(x,v)$
(4)	$C_R \sqsubseteq\, \leq 1 D_A$	$\mathtt{FunctionalDataProperty}(D_A)$	$D_A(x,v_1) \wedge D_A(x,v_2) \Rightarrow$ $v_1 = v_2$
(5)	$\mathsf{ADom} \sqsubseteq\, \leq 1 D_A^{-\,(*)}$	$\mathtt{HasKey}(C_R,(),(D_A))$	$D_A(x_1,v) \wedge D_A(x_2,v) \Rightarrow$ $x_1 \approx x_2$
(6)	$\exists P_A.\top \sqsubseteq C_R$	$\mathtt{ObjectPropertyDomain}(P_A,\ C_R)$	$P_A(x,y) \Rightarrow C_R(x)$
(7)	$\top \sqsubseteq \forall P_A.C_{R'}$	$\mathtt{ObjectPropertyRange}(P_A,\ C_{R'})$	$P_A(x,y) \Rightarrow C_{R'}(y)$
(8)	$C_R \sqsubseteq \exists P_A.C_{R'}$	$\mathtt{SubclassOf}(C_R,$ $\quad\mathtt{ObjectSomeValuesFrom}($ $\quad\quad P_A, C_{R'}))$	$C_R(x) \Rightarrow$ $\exists y.(P_A(x,y) \wedge C_{R'}(y))$
(9)	$P_A \circ D_{A'} \equiv D_A^{(*)}$	$\mathtt{EquivalentDataProperties}($ $\quad\mathtt{DataPropertyChain}(P_A, D_{A'}),$ $\quad D_A)^{(*)}$	$P_A(x,y) \wedge D_{A'}(y,v) \Rightarrow$ $D_A(x,v)$ $D_A(x,v) \Rightarrow$ $\exists y.(P_A(x,y) \wedge D_{A'}(y,v))$
(10)	$C_R \sqsubseteq\, \leq 1 P_A$	$\mathtt{FunctionalObjectProperty}(P_A)$	$P_A(x,y_1) \wedge P_A(x,y_2) \Rightarrow$ $y_1 \approx y_2$

- domain of P_A is C_R (Table 1(6)),
- range of P_A is $C_{R'}$ (Table 1(7)),
- each individual of class C_R has at least one individual connected via P_A (Table 1(8)),
- the data property $P_A \circ D_{A'}$ resulting from the composition of the object property P_A and the data property $D_{A'}$ is equivalent to the data property D_A (Table 1(9)),
- P_A is a functional property (Table 1(10)).

Since $D_{A'}$ is the key for $C_{R'}$ and D_A is a functional property, then it follows from $P_A \circ D_{A'} \equiv D_A$ that P_A must be also a functional property (Table 1(10)).

By the star ($^{(*)}$) we denote expressions illegal in the related notation. In particular, inversions of data properties (concrete roles) are not allowed in standard DL. However, the corresponding feature in the form of $\mathtt{HasKey}$ does exist in OWL 2 EL. Similarly, in standard DL we cannot compose abstract roles with concrete roles. Neither we can perform this operation in OWL 2 EL. However, when the concrete domain (as in our case ADom) is finite and all individuals are named, we can define it and extend the underlying ontological language to define a data property as the composition of an object property from C into C' and a data property from C' into ADom resulting in a data property from C into ADom.

Representation of Relational Database Instance. Let I be an instance of DB. Then the concrete domain ADom is the active domain of the database instance (i.e. consisting of all constants belonging to DB^I) with the natural meaning of equality between constants (i.e. $c_1 = c_2$). The top ($\top$) class consists of the set of individuals isomorphic with the set of all tuples belonging to DB^I. If two tuples are equal, $r = r'$, then the corresponding individuals are equivalent (the same), i.e. $i_r \approx i_{r'}$. Both ADom and $\top$ are then finite.

The instance DB^I is mapped to assertions (facts) in $\mathcal{A}$ as follows (Tab. 2):

Table 2. Representation of database instance

	DL and ORL	OWL 2 EL
(1)	$C_R(i_r)$	$\texttt{ClassAssertion}(C_R, i_r)$
(2)	$D_A(i_r, c)$	$\texttt{DataPropertyAssertion}(D_A, i_r, c)$
(3)	$P_A(i_r, i_{r'})$	$\texttt{ObjectPropertyAssertion}(P_A, i_r, i_{r'})$

1. For each tuple $r \in R^I$, where $R \in \mathbf{R}$, the individual i_r is an instance of the class C_R (Tab. 2(1)).
2. For each constant $c = r.A$, where $A \in att(R)$, $r \in R^I$, $R \in \mathbf{R}$, the data property D_A assigns the individual i_r with the value (literal) c (Tab. 2(2)).
3. Let $fkey(R, A, R', A') \in FKey$ be a foreign key, then for each $r \in R^I$ and $r' \in R'^I$ such that $r.A = r'.A'$, the object property P_A assigns the individual i_r with the individual $i_{r'}$ (Tab. 2(3)).

3.2   Representation of Source-to-Target Dependencies

Let $\mathcal{M} = (DB_s, DB_t, \Sigma_{st})$ be a data exchange setting and $\sigma = \forall \mathbf{u}, \mathbf{v}.(\varphi_s(\mathbf{u}, \mathbf{v}) \Rightarrow \exists \mathbf{w}.\psi_t(\mathbf{u}, \mathbf{w}))$ be a STD in Σ_{st}. Let φ_s be a conjunction of n atomic-formulas over the source database DB_s, and ψ_t be a conjunction of m atomic formulas over the target database DB_t, i.e. $\varphi_s = \alpha_1 \wedge \cdots \wedge \alpha_n$, $\psi_t = \beta_1 \wedge \cdots \wedge \beta_m$.

Let $\mathbf{x} = (x_1, ..., x_n)$ and $\mathbf{y} = (y_1, ..., y_n)$ be two disjoint sequences of different variables. Then the translation $\tau_{(\mathbf{x}.\mathbf{y})}(\sigma)$ of STD σ into conjunction of ORL rules is given by the procedure:

$$\tau_{(\mathbf{x},\mathbf{y})}(\sigma) = \forall \mathbf{u}, \mathbf{v}, \mathbf{x}.(\tau_{\mathbf{x}}(\varphi_s(\mathbf{u}, \mathbf{v})) \Rightarrow \exists \mathbf{w}, \mathbf{y}.\tau_{\mathbf{y}}(\psi_t(\mathbf{u}, \mathbf{w}))),$$
$$\tau_{(x_1,...,x_n)}(\alpha_1 \wedge \cdots \wedge \alpha_n) = \tau_{x_1}(\alpha_1) \wedge \cdots \wedge \tau_{x_n}(\alpha_n),$$
$$\tau_{(y_1,...,y_m)}(\beta_1 \wedge \cdots \wedge \beta_n) = \tau_{y_1}(\beta_1) \wedge \cdots \wedge \tau_{y_n}(\beta_m),$$
$$\tau_x(R(A_1 : v_1, ..., A_k : v_k)) = C_R(x) \wedge D_{A_1}(x, v_1) \wedge \cdots \wedge D_{A_k}(x, v_k),$$
$$\tau_x(v = c) = (v = c),$$
$$\tau_x(v_1 = v_2) = (v_1 = v_2).$$

For example, if $\sigma = \forall u, v.(S(A : u, B : v) \Rightarrow \exists w.T(C : u, D : w))$, then $\tau_{x,y}(\sigma) = \forall u, v, x.(C_S(x) \wedge D_A(x, u) \wedge D_B(x, v) \Rightarrow \exists w, y.C_T(y) \wedge D_C(y, u) \wedge D_D(y, w))$.

4   Semantics Preserving Data Exchange

Let $\mathcal{M} = (DB_s, DB_t, \Sigma_{st})$ be a data exchange setting. Its counterpart involving knowledge bases instead of databases is referred to as *knowledge-based data exchange* setting.

Definition 1. *A knowledge-based data exchange setting corresponding to* $\mathcal{M} = (DB_s, DB_t, \Sigma_{st})$ *is a quadruple* $\mathcal{KM} = (\mathcal{C}_s, \mathcal{C}_t, \mathcal{C}_{st}, \mathcal{S})$, *where:*

- $\mathcal{C}_s = \tau(DB_s)$ - *is the translation of the source database schema* DB_s *into a set of integrity constraint axioms,*
- $\mathcal{C}_t = \tau(DB_t)$ - *is the translation of the target database schema* DB_t *into a set of integrity constraint axioms,*
- $\mathcal{C}_{st} = \tau(\Sigma_{st})$ - *is the translation of the set* Σ_{st} *of STDs over* $\mathbf{R}_s$ *and* $\mathbf{R}_t$,
- $\mathcal{S}$ - *is a set of TBox axioms defining the knowledge about relationships between concepts occurring in the source- and the target knowledge bases.*

The reason for moving from data- to the knowledge-based data exchange is the attempt to control semantics of exchanged data as far as it is possible. Especially interesting are data exchange systems which *preserve semantics of data*, i.e. for which the following three requirements are satisfied:

1. **Admittance of solutions:** for each source instance I there is a universal solution J for I [8], i.e.

$$I \models \mathcal{C}_s \Rightarrow \exists J.(J \models \mathcal{C}_t \wedge (I, J) \models \mathcal{C}_{st}). \tag{1}$$

2. **Consistency with the knowledge base:** if J is a universal solution for I then the pair (I, J) satisfies $\mathcal{S}$, i.e.

$$I \models \mathcal{C}_s \wedge J \models \mathcal{C}_t \wedge (I, J) \models \mathcal{C}_{st} \Rightarrow (I, J) \models \mathcal{S}. \tag{2}$$

3. **Maximal generality:** if the two conditions above are satisfied for an instance I and its universal solution J, then they are not satisfied for any proper superset I' of I and its solution J', i.e.

$$\begin{aligned} I \models \mathcal{C}_s \wedge J \models \mathcal{C}_t \wedge (I, J) \models \mathcal{C}_{st} \wedge (I, J) \models \mathcal{S} \wedge \\ I' \models \mathcal{C}_s \wedge J' \models \mathcal{C}_t \wedge (I', J') \models \mathcal{C}_{st} \wedge (I', J') \models \mathcal{S} \Rightarrow \neg(I \subset I'). \end{aligned} \tag{3}$$

The following example illustrates the problem of semantics preservation in data exchange systems. Let us consider two relational database schemas:

$$\begin{aligned} DB_s = (\ \mathbf{R}_s = \ &\{\mathtt{s:Student(SId, Name, Type)}, \mathtt{s:Exam(EId, ESId, Course, Grade)}\}, \\ IC_s = \ &\{pkey(\mathtt{s:Student, SId}), pkey(\mathtt{s:Exam, EId}), \\ &\quad fkey(\mathtt{s:Exam, ESId, s:Student, SId})\}; \\ DB_t = (\ \mathbf{R}_t = \ &\{\mathtt{t:Student(SId, Name)}, \mathtt{t:Exam(EId, ESId, Course, Grade)}\}, \\ IC_t = \ &\{pkey(\mathtt{t:Student, SId}), pkey(\mathtt{t:Exam, EId}), \\ &\quad fkey(\mathtt{t:Exam, ESId, t:Student, SId})\}. \end{aligned}$$

The source database DB_s stores data about students (undergraduate and graduate) and their exams, while the target data base (DB_t) stores the data only about graduate students. The Type attribute takes "g" for graduate, and "u" for undergraduate students. Each student is identified by student identifier (SId), and have a name (Name). Each exam is identified be exam identifier (EId), and represents the fact that the student identified by ESId passed exam in Course getting the Grade.

Further on, ontological concepts or properties corresponding to symbols in database schemas will be denoted by the same name but using italic. For example, the class $C_{\text{t:Student}}$ is written as $t{:}Student$, and the data property D_{Type} assigned to the attribute Type from DB_s, is written as $s{:}Type$. Prefixes $s{:}$ or $t{:}$ are used to denote that the corresponding concept belongs to the ontology related to, respectively, the source- or target database schema.

The following set of axioms in DL describes interrelationships between concepts from the source- and the target knowledge bases.

$$\mathcal{S} = \{t{:}Student \sqsubseteq s{:}Student, t{:}Exam \sqsubseteq s{:}Exam, \exists(s{:}Type).\{"g"\} \equiv t{:}Student\}.$$

This set can be translated into the following set of ORL rules (note that classes are represented by their key data properties):

$$\mathcal{S} = \{ \ t{:}SId(y,v) \Rightarrow \exists x.(s{:}SId(x,v)), t{:}EId(y,v) \Rightarrow \exists x.(s{:}EId(x,v)),$$
$$s{:}SId(x,v) \wedge s{:}Type(x,w) \wedge w = "g" \Rightarrow \exists y.(t{:}SId(y,v)),$$
$$t{:}SId(y,v) \Rightarrow \exists x.(s{:}SId(x,v) \wedge s{:}Type(x,w) \wedge w = "g")\}.$$

The set $\mathcal{C}_{st}$ of STDs, defining mappings between the source and the target knowledge bases, will be verified against the three requirements underlying the semantics preserved data exchange (using the closed world assumption [16]).

1. Let $\Sigma_{st} = \{\sigma\}$, where $\sigma = \text{s}{:}\text{Student}(v_1, v_2, v_3) \Rightarrow \text{t}{:}\text{Student}(v_1, v_2)$. Then $s{:}SId(x,v) \Rightarrow \exists y.(t{:}SId(y,v)) \in \mathcal{C}_{st}$ And we see that the set

$$\{t{:}SId(y,v) \Rightarrow \exists x.(s{:}SId(x,v)), \ s{:}SId(x,v) \Rightarrow \exists y.(t{:}SId(y,v))\}$$

violates the consistency requirements (2). The set would be consistent if $s{:}Student$ and $t{:}Student$ were equivalent. But this is not the case.

2. Let $\Sigma_{st} = \{\sigma_1, \sigma_2\}$, where

$$\sigma_1 = \text{s}{:}\text{Student}(v_1, v_2, v_3) \wedge v_3 = "g" \Rightarrow \text{t}{:}\text{Student}(v_1, v_2),$$
$$\sigma_2 = \text{s}{:}\text{Exam}(w_1, w_2, w_3, w_4) \Rightarrow \text{t}{:}\text{Exam}(w_1, w_2, w_3, w_4).$$

We see that $\tau(\sigma_1)$ is consistent with $\mathcal{S} \cup \mathcal{C}_t$, but σ_2 is not. Moreover, σ_2 violates the foreign key integrity constraint in the target database for the students having in the source database the value of $s{:}Type$ different from "g". Indeed, let x be such $s{:}Student$ individual (tuple) that $s{:}SId(x, w_2) \wedge s{:}Type(x,w) \wedge w = "u"$. Then:

$$s{:}EId(x, w_1) \wedge s{:}ESId(x, w_2) \Rightarrow \exists y.(t{:}EId(y, w_1) \wedge t{:}ESId(y, w_2)),$$
$$t{:}ESId(y, w_2) \Rightarrow \exists z.(t{:}SId(z, w_2)),$$
$$t{:}SId(z, w_2) \Rightarrow \exists r.(s{:}SId(r, w_2) \wedge s{:}Type(r, w) \wedge w = "g").$$

and we have contradiction with the assumption. Thus, the admittance requirement (1) is not satisfied.

3. Let $\Sigma_{st} = \{\sigma_3\}$, where

$$\sigma_3 = \mathsf{s:Student}(v_1, v_2, v_3) \wedge v_3 = "g" \wedge \mathsf{s:Exam}(w_1, v_2, w_3, w_4) \Rightarrow$$
$$\mathsf{t:Student}(v_1, v_2, v_3) \wedge \mathsf{t:Exam}(w_1, v_2, w_3, w_4).$$

It is easy to show that $\mathcal{S} \cup \mathcal{C}_t \cup \tau(\sigma_3)$ is consistent. However, we can see that some data, that potentially could be mapped, are not mapped by σ_3. In particular, graduate students who did not pass any exam, do not participate in the mapping (do not satisfy the left hand side of σ_3). To be more precise, note that σ_3 can be decomposed into two dependencies:

$$\sigma_3' = \mathsf{s:Student}(v_1, v_2, v_3) \wedge v_3 = "g" \wedge \mathsf{s:Exam}(w_1, v_2, w_3, w_4) \Rightarrow$$
$$\mathsf{t:Student}(v_1, v_2, v_3),$$
$$\sigma_3'' = \mathsf{s:Student}(v_1, v_2, v_3) \wedge v_3 = "g" \wedge \mathsf{s:Exam}(w_1, v_2, w_3, w_4) \Rightarrow$$
$$\mathsf{t:Exam}(w_1, v_2, w_3, w_4).$$

We see that σ_3'' is formulated correctly – it maps as many data as expected. In contrast, σ_3' does not map all expected data. For example, for the instance

$$I = \{ \mathsf{s:Student}(st1, joe, g), \mathsf{s:Student}(st2, eve, g), \mathsf{s:Student}(st3, ann, u),$$
$$\mathsf{s:Exam}(ex1, st1, math, A), \mathsf{s:Exam}(ex2, st3, dbs, B)\}.$$

the fact $\mathsf{s:Student}(st2, eve, g)$ will not be mapped by σ_3'. However, if the last conjunct on the left hand side of σ_3' is dropped, then we obtain

$$\sigma_4 = \mathsf{s:Student}(v_1, v_2, v_3) \wedge v_3 = "g" \Rightarrow \mathsf{t:Student}(v_1, v_2, v_3),$$

and σ_4 maps the data as expected. Moreover, σ_4 remains still consistent with $\mathcal{S} \cup \mathcal{C}_t$.

4. To summarize, we see that $\Sigma = \{\sigma_4, \sigma_3''\}$ satisfies all requirements concerning semantics preserving data exchange.

5   Conclusion

In this paper we study the problem of reasoning about schema mappings between two database schemas DB_s and DB_t in data exchange systems. This problem is of significant importance in many applications, especially in data integration systems. The approach consists in: (1) Translating source- and target database schemas into sets $\mathcal{C}_s$ and $\mathcal{C}_t$ of integrity constraint axioms, and creating a set $\mathcal{S}$ of general knowledge axioms concerning the application domain. We show that this can be achieved using the expressive power of OWL 2 EL ontology (with a slight extension). (2) Creation of a knowledge-based data exchange system with a set $\mathcal{C}_{st}$ of source-to-target dependencies describing how the data from the source database are to be transformed into the target database. (3) The crucial issue is the control of semantics preservation during transformations in the data exchange. We extended the requirement (admittance) formulated against the data-oriented data exchange systems, adding requirements about consistency with the S axioms and maximal generality. We identified this problem by developing the SixP2P system for semantic integration of data in P2P environment [6,14], where the described method constitutes fundamentals for verifying mappings between P2P-connected databases.

References

1. Abiteboul, S., Hull, R., Vianu, V.: Foundations of Databases. Addison-Wesley, Reading (1995)
2. Baader, F., Calvanese, D., McGuinness, D., Nardi, D., Petel-Schneider, P. (eds.): The Description Logic Handbook: Theory, Implementation and Applications. Cambridge University Press (2003)
3. Beneventano, D., Bergamaschi, S.: The MOMIS methodology for integrating heterogeneous data sources. In: IFIP Congress Topical Sessions, pp. 19–24 (2004)
4. Bernstein, P.A., Haas, L.M.: Information integration in the enterprise. Commun. ACM 51(9), 72–79 (2008)
5. Bonifati, A., et al.: Schema mapping and query translation in heterogeneous P2P XML databases. VLDB J. 19(2), 231–256 (2010)
6. Brzykcy, G., Bartoszek, J., Pankowski, T.: Schema Mappings and Agents' Actions in P2P Data Integration System. Journal of Universal Computer Science 14(7), 1048–1060 (2008)
7. Calvanese, D., De Giacomo, G., Lembo, D., Lenzerini, M., Rosati, R., Ruzzi, M.: Using OWL in Data Integration. In: Semantic Web Information Management, ch. 17, pp. 397–424. Springer (2010)
8. ten Cate, B., Kolaitis, P.G.: Structural characterizations of schema-mapping languages. Commun. ACM 53(1), 101–110 (2010)
9. Cruz, I.F., Xiao, H.: Ontology Driven Data Integration in Heterogeneous Networks. In: Complex Systems in Knowledge-based Environments, pp. 75–98 (2009)
10. Fagin, R., Kolaitis, P.G., Miller, R.J., Popa, L.: Data exchange: semantics and query answering. Theor. Comput. Sci. 336(1), 89–124 (2005)
11. Grosof, B.N., Horrocks, I., Volz, R., Decker, S.: Description logic programs: combining logic programs with description logic. In: WWW, pp. 48–57 (2003)
12. Motik, B., Horrocks, I., Sattler, U.: Bridging the gap between OWL and relational databases. Journal of Web Semantics 7(2), 74–89 (2009)
13. OWL 2 Web Ontology Language Profiles (2009),
   http://www.w3.org/TR/owl2-profiles
14. Pankowski, T.: Query propagation in a P2P data integration system in the presence of schema constraints. In: Hameurlain, A. (ed.) Globe 2008. LNCS, vol. 5187, pp. 46–57. Springer, Heidelberg (2008)
15. Pankowski, T.: Using Data-to-Knowledge Exchange for Transforming Relational Databases to Knowledge Bases. In: Bikakis, A., Giurca, A. (eds.) RuleML 2012. LNCS, vol. 7438, pp. 256–263. Springer, Heidelberg (2012)
16. Reiter, R.: On Closed World Data Bases. Logic and Data Bases, 55–76 (1977)
17. Sequeda, J., Tirmizi, S.H., Corcho, Ó., Miranker, D.P.: Survey of directly mapping SQL databases to the Semantic Web. Knowledge Eng. Review 26(4), 445–486 (2011)
18. Xiao, H., Cruz, I.F.: Integrating and Exchanging XML Data Using Ontologies. In: Spaccapietra, S., Aberer, K., Cudré-Mauroux, P. (eds.) Journal on Data Semantics VI. LNCS, vol. 4090, pp. 67–89. Springer, Heidelberg (2006)

Multi-Relational Learning for Recommendation of Matches between Semantic Structures

Andrzej Szwabe, Pawel Misiorek, and Przemyslaw Walkowiak

Institute of Control and Information Engineering, Poznan University of Technology,
M. Sklodowskiej-Curie Square 5, 60-965 Poznan, Poland
{Andrzej.Szwabe,Pawel.Misiorek,Przemyslaw.Walkowiak}@put.poznan.pl

Abstract. The paper presents the Tensor-based Reflective Relational Learning System (TRRLS) as a tensor-based approach to automatic recommendation of matches between nodes of semantic structures. The system may be seen as realizing a probabilistic inference with regard to the relation representing the 'semantic equivalence' of ontology classes. Despite the fact that TRRLS is based on the new idea of algebraic modeling of multi-relational data, it provides results that are comparable to those achieved by the leading solutions of the Ontology Alignment Evaluation Initiative (OAEI) contest realizing the task of matching concepts of Anatomy track ontologies on the basis of partially known expert matches.

Keywords: statistical relational learning, tensor-based data modeling, ontology matches recommendation, RDF-based reasoning.

1   Introduction

The Encyclopedia of Machine Learning [12] presents Statistical Relational Learning (SRL) as the approach targeting the main goals of Artificial Intelligence. In recent years, tensor-based approaches to SLR have become a promising alternative to commonly used relational learning methods based on graphical models [3], [7], [10], [13].

The probabilistic modeling of SRL methods based on graphical models enhances the results of structure learning by representing probabilistically the set of first-order logic propositions explicitly provided to the system [4]. In contrast to such methods, an SRL method that is based on algebraic data representations (such as the one presented in this paper) imposes no arbitrary assumptions with regard to the model's structure, i.e., no arbitrary distinction between the phases of structure learning and parameter learning is necessary. It should be admitted that assumptions that are made with regard to a graphical model structure are not necessarily 'arbitrary', but only as long we assume that expert knowledge is more reliable source of knowledge about the domain than patterns of coincidences appearing in the input data.

The 3rd-order tensor is known as a data structure that allows for very convenient representation of heterogeneous relational data, including data provided

M. Graña et al. (Eds.): KES 2012, LNAI 7828, pp. 98–107, 2013.

as RDF triples [6], [10]. The Tensor-based Reflective Relational Learning System (TRRLS) presented in this paper is based on one of the first tensor-based Statistical Relational Learning solutions – referred to as Tensor-based Reflective Relational Learning Framework (TRRLF). One of the key components of TRRLF is a probabilistic model of the semantic structure representation consistency. The probabilistic modeling emphasized in the paper is based on the probabilistic vector-space similarity modeling. We propose to extend the pair-oriented IR-derived coincidence probability model into a triple-oriented model what implied the need for a redefinition of the probabilistically interpretable inner product. A rigorous modeling of unambiguously interpretable probabilities corresponding to all logical propositions that a TRRL system deals with, enables its fully automatic operation.

The TRRL framework may be seen as a new approach combining tensor-based relational learning, probabilistic modeling, and reflective data processing. This way our solution may be regarded as a new contribution to SRL for which the tensor-based approach has been already recognized as very promising [12]. To our knowledge, TRRLS is the first SRL system that is based on the probabilistic modeling and introduces elements of the Reflective Random Indexing (RRI) technique (a highly-effective data processing method established in the area of IR [2]).

2 Related Work

The Encyclopedia of Machine Learning [12] presents tensor-based data modeling as a research trend initiated by the research on collective matrix factorization applied to relational data [11]. Sutskever et al. [13] employed the tensor factorization approach to the framework aimed at inferencing from relational data. The model is based on Bayesian clustering and is not designed to deal with data provided as RDF triples. Recent works on applying tensors to represent relational data given as RDF triples include [6] and [10]. Franz et al. [6] used the PARAFAC decomposition as a means for tensor-based semantic graph representation generation. The approach presented by Nickel et al. [10] contributes with an efficient algorithm to compute the factorization of a 3rd-order tensor, but it is limited to the dyadic relational data model.

None of these approaches targets the probabilistic interpretation of input data and processing results. The existing tensor-based relational data processing frameworks are founded on different applications of tensor factorization [8], rather than on theoretically-grounded probabilistic modeling. In contrast to the existing tensor-based approaches to SRL [6], [10], we model propositional data in a way that enables full flexibility of specifying the roles that any pair of entities plays with regard to any relation. Instead of using the dyadic relational model [10], we represent the 'active mode' (corresponding to the relation's subject) and the 'passive mode' (as the relation's object) 'views' of a given entity as potentially fully independent to each other.

In contrast to other tensor-based approaches to SRL [13], [6], [10], TRRL does not involve the application of the heuristic dimensionality reduction based on some factorization of matrix or tensor data. Although in our method all subjects, predicates, and objects are represented in a space of d-dimensions, the method itself does not force us to apply a dimensionality reduction of any form: when long computation time is acceptable, we may use initial vectors that are fully orthogonal instead of employing random indexing. Thus, the dimensionality reduction is not employed to make our system work or to improve its performance (by reducing 'noise', such it is in the case of Latent Semantic Analysis or similar methods) but for purely practical reasons of computational scalability. While the approaches to SRL founded on tensor data representations resemble legacy Natural Language Processing (NLP) methods such as LSA [6], [10], TRRL is much more strongly related to similarity-driven reflective data processing methods, such as RRI. It is worth noting that, at least with respect to the fields of IR and NLP, reflective data processing methods are regarded by some authors as outperforming more methods based on the factorization-based dimensionality reduction heuristics [1].

3   Multi-relational Learning System

3.1   Tensor-Based Propositional Data Representation

Let us introduce system T as representing subject-predicate-object dependencies in the form of a 3rd-order tensor (understood as a multidimensional array). System T models the set of relations (predicates) $R = \{r_k\}$ between subjects from set $S = \{s_i\}$ and the objects from set $O = \{o_j\}$. We assume that $|R| = m$ and that $|S| = |O| = n$. The latter assumption is motivated by the fact that we allow each entity, that we would like to represent in the tensor, to play the role of a subject or an object. We define the tensor as $T_{i,j,k} = [t_{i,j,k}]_{n \times n \times m}$, where value $t_{i,j,k}$ represents our initial knowledge about relation k from subject i to object j. We assume that $t_{i,j,k} \geq 0$ for every i, j, k. Additionally, we define set E as a set of all elements described in the system ($E = S \cup O \cup R$), and set F as a set of all known facts (i.e., RDF triples (i, j, k)) which are used to build the input tensor. The number $|F| = f$ determines the number of positive cells in the input tensor.

3.2   Vector-Space Representation of Heterogeneous Entities

An additional representation (called the context vector) in the d-dimensional vector space has to be provided for each entity described in T, i.e., for relations, objects, and subjects, as well as for all known facts stored in the input tensor. Context vectors are stored in matrix $X = [x_{i,j}]_{(2n+m+f) \times d}$. The content of this matrix determines the results of training, which is performed to explore the correspondences between entities from set $E = S \cup O \cup R$, based on the information about 'connections' between them stored in the tensor. We assume that the first n rows of matrix X represent items from set S, the rows indexed from $n + 1$ to $2n$ represent items from set O, and the next rows indexed from

$2n + 1$ to $m + 2n$ represent relations from set R. We refer the top n rows of X to as $n \times d$ matrix X^S (of vectors x_i^S for $i = 1..n$), the next n rows (rows indexed from $n + 1$ to $2n$) to as $n \times d$ matrix X^O (of vectors x_j^O for $j = 1..n$), and the next m rows to as $m \times d$ matrix X^R (of vectors x_k^R for $k = 1..m$). Additionally, we denote the top $2n + m$ rows of X by $(2n + m) \times d$ matrix X^E representing context vectors of all the elements from set $E = S \cup O \cup R$. The bottom f rows of matrix X contain context vectors representing facts from the set F and form $f \times d$ matrix X^F.

Reflective learning on the basis of the connections between entities modeled as a 3-rd order tensor is enabled by using the common space for all context vectors. Such a representation makes it possible to model all the real-world objects described in the system as compatible objects [15]. The TRRL framework allows to configure the length of context vectors by setting parameter d.

3.3 Proposition Probability Space

We introduce the probability space that is based on events $A_{i,j,k}$ corresponding to the presence of relation k $(r_k \in R)$ from subject i $(s_i \in S)$ to object j $(o_j \in O)$. The probability measure is defined according to a distribution which can be presented as the 3-rd order tensor $T^P = [t_{i,j,k}^P]_{n \times n \times m}$, where $t_{i,j,k}^P = P(A_{i,j,k})$ for $i = 1..n$, $j = 1..n$, $k = 1..m$.

The input tensor is built from RDF triples, in such a way that each tensor cell (i, j, k), which corresponds to some RDF triple from set F (i.e., for which we know from input data that relation k for subject i and object j holds), is equal to the same positive value, whereas all the values in the remaining cells are set to be equal 0. Then, the tensor is normalized in order to obtain the distribution $t_{i,j,k}^P$ (i.e., in order to ensure that $\sum_{i,j,k} t_{i,j,k}^P = 1$).

The probabilities for particular elements described by tensor T (subjects from S, objects from O, and relations from R), i.e., probabilities of events:

- A_i^S - the event that some relation from subject i to some object is observed,
- A_j^O - the event that some relation from some subject to object j is observed,
- A_k^R - the event that relation k from some subject to some object is observed,

are calculated based on distribution described by T^P; it is done by adding entries of the tensor slice corresponding to a given element. In particular, we have $P(A_i^S) = \sum_{j,k} t_{i,j,k}^P$ for $i = 1..n$, $P(A_j^O) = \sum_{i,k} t_{i,j,k}^P$ for $j = 1..n$, and $P(A_k^R) = \sum_{i,j} t_{i,j,k}^P$ for $k = 1..m$. These probabilities are used in the procedure of tensor reconstruction that is executed after the learning phase.

The way of building the probability distribution $T_{i,j,k}^P$ implies that the probabilities of events A_i^S, A_j^O, and A_k^R are different, in dependence from the characteristics of the input data. Following the principle of indifference, for the purposes of learning and matches calculation we use conditional events $A_{i,j,k}|A_i^S$,

$A_{i,j,k}|A_j^O$, and $A_{i,j,k}|A_k^R$. We assume that events $A_{i,j,k}|A_i^S$, $A_{i,j,k}|A_j^O$, and $A_{i,j,k}|A_k^R$ are independent. More precisely, we build the tensor $T' = [t'_{i,j,k}]_{n \times n \times m}$, where

$$
\begin{aligned}
t'_{i,j,k} &= P\left((A_{i,j,k}|A_i^S) \cap (A_{i,j,k}|A_j^O) \cap (A_{i,j,k}|A_k^R)\right) \\
&= P\left(A_{i,j,k}|A_i^S\right) P\left(A_{i,j,k}|A_j^O\right) P\left(A_{i,j,k}|A_k^R\right) \\
&= \frac{P(A_{i,j,k})}{P(A_i^S)} \frac{P(A_{i,j,k})}{P(A_j^O)} \frac{P(A_{i,j,k})}{P(A_k^R)}.
\end{aligned}
\tag{1}
$$

Probabilistic relevance modeling applied to quantum information retrieval methods [15] is the basis for the reconstruction of the TRRLF tensor from the set of context vectors. It is worth stressing that our method does not involve a typical tensor factorization – a cell is not factorized into a set of additive components, but is reconstructed, using the formula derived from quantum probability calculations. As a result, reconstructed tensor $\widehat{T} = [\widehat{t}_{i,j,k}]_{n \times n \times m}$ is built:

$$
\widehat{t}_{i,j,k} = cos^2(x_i^S, x_j^O)cos^2(x_i^S, x_k^R)cos^2(x_j^O, x_k^R).
\tag{2}
$$

3.4 Multi-relational Learning Procedure

The learning procedure uses both tensor T^P and matrix X. Matrix X consists of context vectors which are updated during learning, whereas the tensor is applied as a source of data used in the learning process.

We propose the 3rd-order Tensor-based Reflective Indexing (3-TRI) procedure, which allows each vector to be 'learned' by other context vectors according to the 'connections' between entities represented in tensor T^P. Similarly to the method presented in [14], the 3-TRI procedure presented in this paper is based on an application of consecutive reflections in a way similar to the RRI approach [2]. The tensor T^P is used to construct matrix $Y = [y_{i,j}]_{(2n+m) \times f}$ which describes correlations between entities and facts represented as tensor T^P. In particular, matrix Y is calculated in the following way:

$$
y_{i,j} = \begin{cases} \sqrt{\frac{1}{n_i}} & \text{if element } i \text{ is the subject, object or predicate of fact } j, \\ 0 & \text{otherwise,} \end{cases}
\tag{3}
$$

for $1 \leq i \leq 2n + m$, and $1 \leq j \leq f$, where n_i is the number of facts in F, which concern element i. As follows from the above formulas the matrix Y is a sparse matrix: each its column contains only three nonzero values.

At the first step of the 3-TRI procedure, for each row of matrix X^F we select s coordinates (uniformly at random from the set of d dimensions) and set them to be equal to 1. Then each of rows of X^F is normalized using the formula:

$$
x_{i,\cdot}^F = \frac{x_{i,\cdot}^F}{\|x_{i,\cdot}^F\|_2}, \text{ for } 1 \leq i \leq f.
\tag{4}
$$

Each reflection step of the 3-TRI procedure is based on the following context vector update procedure (Eq. (5-10)):

$$X^E := YX^F \tag{5}$$

$$\forall_{i=1..2n+m}\forall_{j=1..d} \ x^E_{i,j} := \sqrt{x^E_{i,j}} \tag{6}$$

$$x^E_{i,\cdot} := \frac{x^E_{i,\cdot}}{\|x^E_{i,\cdot}\|_2}, \text{ for } 1 \leq i \leq 2n + m \tag{7}$$

$$X^F := Y^T X^E \tag{8}$$

$$\forall_{i=1..f}\forall_{j=1..d} \ x^F_{i,j} := \sqrt{x^F_{i,j}} \tag{9}$$

$$x^F_{i,\cdot} := \frac{x^F_{i,\cdot}}{\|x^F_{i,\cdot}\|_2}, \text{ for } 1 \leq i \leq f \tag{10}$$

After each reflection step, for each known fact (i, j, k) from set F, we calculate its value using the reconstruction formula (Eq. (2)). The learning procedure stops when all the reconstructed values $\hat{t}_{i,j,k}$ are at least as big as corresponding values $t'_{i,j,k}$ (calculated according to Eq. (1)).

4 Application in an Ontology Alignment Scenario

The TRRL system has been applied for the OAEI Anatomy track [16]. According to the rules of OAEI, TRRLS has used two ontologies as an input (a source and a target one). Input OWL documents have been represented as a set of matrices, where each matrix corresponds to one of the relations.

Figure 1 presents the way the tensor for the OAEI Anatomy track scenario is constructed. Matrices of relations are the first slices of the tensor: r_1, r_2,...., r_n. Each slice representing a relation consists of two submatrices describing the relation for the source ontology and the target ontology, respectively. The values of these submatrices indicate the presence of the relation between entities from a given ontology. The next *hasTerm* and *termOf* slices represent terms that 'describe' the entities. The use of the *termsOf* slice is optional. This relation is the inverse of the *hasTerm* relation. The *weakIdentity* slice is a diagonal matrix combining the role of an object and the role of a subject for the same entity. The final correspondences between entities are determined using the slice *matchesTo*, which is built using the information about partial matches that is available to the OAEI competitors for the purposes of Subtask #4 of the OAEI Anatomy track [5]. The structure-centric approach of TRRL, which is typical for the SRL methods, is clearly visible in the relational representation of all the data provided to a system: terms occurring in the names of ontology classes are represented

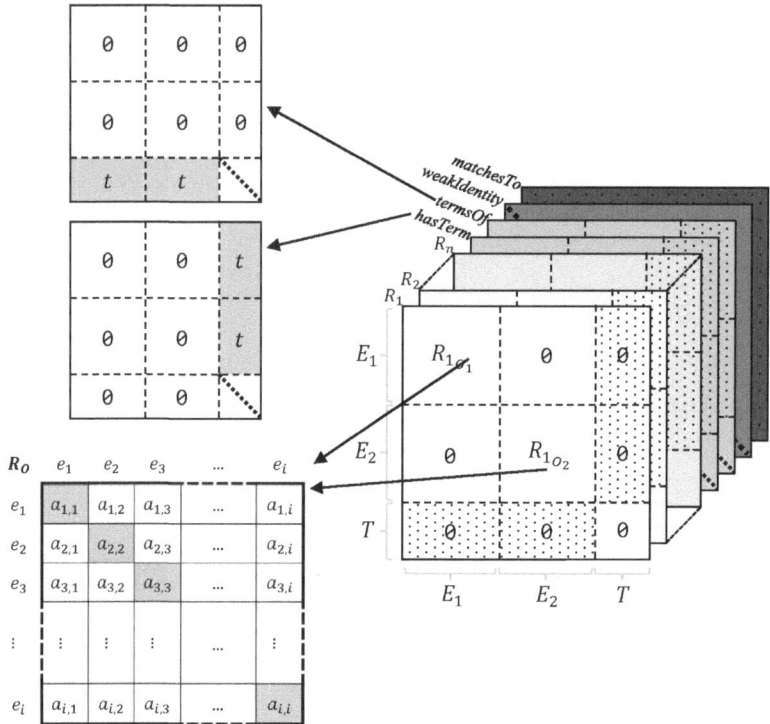

Fig. 1. The view on the TRRL tensor structure

just like entities (TRRL 'objects'), being in the *hasTerm* relation with entities representing the classes.

The procedure of matches generation follows the context vector update phase, described in Subsection 3.4. In order to calculate the tensor values a reconstruction formula (Eq. (2)) is used, where x_i^S corresponds to entities from compared ontologies in the subject mode, x_j^O corresponds to entities from compared ontologies in the object mode, and x_k^R corresponds to the *matchesTo* relation. The submatrix obtained this way is used to establish the final matches. The matches selection procedure is based on selecting maximum values over the rows and columns and then applying the thresholding operation.

5 Experimental Evaluation

The experiments presented in this section were performed using the OAEI 2010 dataset from the Anatomy track [5] in the scenario of identifying matches between the concepts of Adult Mouse Anatomy (MA) and NCI Thesaurus ontologies. Instead of providing the results for the basic OAEI Anatomy Subtask #1,

Table 1. Selected results of Anatomy Subtask #4 track of OAEI 2010

Competitor	Recall	Precision	F-measure
AgreementMaker	0.851	0.929	0.888
ASMOV	0.808	0.837	0.822
CODI	0.746	0.968	0.843

we investigate the scenario of OAEI Anatomy Subtask #4, which is focused on the discovery of nontrivial matches and involves using the data set that includes the so-called partial matches [16]. The hybrid dataset that we have used contains all the literal (i.e., trivial) matches complemented by the set of 54 non-literal matches. The main reason for such a choice is the fact that we investigate osten-sive retrieval [15] – i.e., we assume the use of behavioral data (i.e. the *matchesTo* relation) while defining the semantics of non-behavioral relations [9]. Subtask #4 allows us to introduce the 'behavioral dimension' of the *matchesTo* relation in our method of matches generation.

We have compared results obtained while using our method with those provided by the systems competing in OAEI 2010 Anatomy Subtask #4 [5]: the Agreement Maker (AgrMaker), ASMOV - Automated Semantic Mapping of Ontologies with Validation (ASMOV), and Combinatorial Optimization for Data Integration Sys-tem (CODI). In order to enable the comparison, we have used the performance measures used in OAEI contest, i.e., Precision (P), Recall (R), and F-measure (F) [5]. It is worth to note that in the case of scenarios other than the ones based on the OAEI datasets, the system performance evaluation may be enhanced by the use of measures not limited to the Precision, Recall, and F-measure - e.g., the AuROC (Area under Receiver Operating Characteristic) curve.

Table 2. The experimentation results

	TRRLS(800)			TRRLS(1600)		
	P	**R**	**F**	**P**	**R**	**F**
avg	0.901	0.765	0.827	0.908	0.770	0.833
stddev	0.002	0.002	0.002	0.002	0.002	0.002
min	0.897	0.762	0.824	0.904	0.768	0.831
max	0.905	0.768	0.831	0.911	0.773	0.836

In the case of experiments presented in this paper, the TRRL system has produced a tensor representing 2737 entities from MA ontology, 3298 entities for NCI ontology, 2193 terms, and the total number of 53427 facts. We compare the performance of the system for two different settings of context vector space dimensionality: for $d = 800$ (TRRLS800), and for $d = 1600$ (TRRLS1600). We use seed $s = 2$ (the number of non-zero entries per vector) for the random ini-tialization of matrix X^F. Moreover, we set the threshold value for the matches generation procedure in such a way that the system is allowed to provide addi-tional 300 matches more than the available 987 partial matches.

In order to obtain the results which are comparable with those obtained by the competitors of OAEI 2010 Anatomy Subtask #4, we strictly follow the methodology described in [5]. Since we use a random method to generate the initial form of context vector matrix, we have repeated each experiment 10 times in order to evaluate the impact of the method's randomness on the matching quality. Table 2 presents the Precision, Recall, and F-measure results for TRRLS800 and TRRLS1600 in terms of the minimum value, the maximum value, the average value, and the standard deviation.

The results of Subtask #4 of OAEI 2010 Anatomy track are presented in Table 1. AgrMaker and ASMOV provide relatively good recall, but these systems use external knowledge sources (they use WordNet and UMLS) in order to increase their performance. CODI is based on lexical similarity measures combined with the Markov logic approach. The system has relatively good precision but suffers from relatively weak recall. In contrast, our method is neither supported by an external knowledge base nor focuses on the lexical similarity of matched nodes names. Despite those, quite severely limiting assumptions, the TRRL system allows to achieve matching quality comparable the state-of-the-art methods.

6   Conclusions

The Tensor-based Reflective Relational Learning System presented in the paper may be regarded as a general-purpose semantic structure matching tool. We have demonstrated that, despite the fact that the tool is fully self-configurable and does not utilize external knowledge sources, it may be successfully applied to the ontology alignment task. We believe that the future system improvements (e.g., aimed at introduction of domain knowledge) will lead to even better performance of TRRLS.

TRRL is a fairly new approach to SRL, so its practical value in various application areas has yet to be evaluated. Being an SRL solution, the TRRL framework is likely to be successfully applicable to systems that recommend semantic schema matches, as well as to AI systems from other application areas, in which case the probability, rather than the provability of automated inferences, is the key issue.

Acknowledgments. This work is supported by the Polish Ministry of Science and Higher Education under grant N N516 196737 and by Polish National Science Centre under grant DEC-2011/01/D/ST6/06788.

References

1. Ciesielczyk, M., Szwabe, A.: RI-based Dimensionality Reduction for Recommender Systems. In: Proc. of 3rd International Conference on Machine Learning and Computing. IEEE Press, Singapore (2011)
2. Cohen, T., Schaneveldt, R., Widdows, D.: Reflective Random Indexing and Indirect Inference: A Scalable Method for Discovery of Implicit Connections. Journal of Biomedical Informatics 43(2), 240–256 (2010)

3. De Raedt, L.: Logical and Relational Learning. Springer (2008)
4. Dietterich, T., Domingos, P., Getoor, L., Muggleton, S., Tadepalli, P.: Structured Machine Learning: the Next Ten Years. Machine Learning 73(1), 3–23 (2008)
5. Euzenat, J., Ferrara, A., Meilicke, C., Nikolov, A., Pane, J., Scharffe, F., Shvaiko, P., Stuckenschmidt, H., Svb-Zamazal, O., Svtek, V., Trojahn dos Santos, C.: Results of the Ontology Alignment Evaluation Initiative 2010. In: Proc. of 5th ISWC Workshop on Ontology Matching (OM), Shanghai, pp. 85–117 (2010)
6. Franz, T., Schultz, A., Sizov, S., Staab, S.: TripleRank: Ranking Semantic Web Data by Tensor Decomposition. In: Bernstein, A., Karger, D.R., Heath, T., Feigenbaum, L., Maynard, D., Motta, E., Thirunarayan, K. (eds.) ISWC 2009. LNCS, vol. 5823, pp. 213–228. Springer, Heidelberg (2009)
7. Getoor, L., Taskar, B.: Introduction to Statistical Relational Learning. The MIT Press (2007)
8. Kolda, T.G., Bader, B.W.: Tensor Decompositions and Applications. SIAM Review 51(3), 455–500 (2009)
9. Lavrenko, V.: A Generative Theory of Relevance. Springer, Berlin (2010)
10. Nickel, M., Tresp, V., Kriegel, H.-P.: A Three-Way Model for Collective Learning on Multi-Relational Data. In: Proceedings of the 28th International Conference on Machine Learning (2011)
11. Singh, A.P., Gordon, G.J.: Relational Learning via Collective Matrix Factorization. In: Proceeding of the 14th ACM SIGKDD International Conference on Knowledge Discovery and Data Mining, pp. 650–658 (2008)
12. Struyf, J., Blockeel, H.: Relational Learning. In: Sammut, C., Webb, G. (eds.) Encyclopedia of Machine Learning, pp. 851–857. Springer (2010)
13. Sutskever, I., Salakhutdinov, R., Tenenbaum, J.B.: Modelling Relational Data Using Bayesian Clustered Tensor Factorization. Advances in Neural Information Processing Systems 22 (2009)
14. Szwabe, A., Ciesielczyk, M., Misiorek, P.: Long-tail Recommendation Based on Reflective Indexing. In: Wang, D., Reynolds, M. (eds.) AI 2011. LNCS, vol. 7106, pp. 142–151. Springer, Heidelberg (2011)
15. van Rijsbergen, C.J.: The Geometry of Information Retrieval. Cambridge University Press, New York (2004)
16. Ontology Alignment Evaluation Initiative. 2011 Campaign (2011), http://oaei.ontologymatching.org/2011/

Semantically Enhanced Text Stemmer (SETS) for Cross-Domain Document Clustering

Ivan Stankov, Diman Todorov, and Rossitza Setchi

Knowledge Engineering Systems Group,
School of Engineering, Cardiff University, UK
{stankovid,todorovd,setchi}@cardiff.ac.uk

Abstract. This paper focuses on processing cross-domain document repositories, which is challenged by the word ambiguity and the fact that monosemic
words are more domain-oriented than polysemic ones. The paper describes a
semantically enhanced text normalization algorithm (SETS) aimed at improving
document clustering and investigates the performance of the sk-means clustering algorithm across domains by comparing the cluster coherence produced
with semantic-based and traditional (TF-IDF-based) document representations.
The evaluation is conducted on 20 generic sub-domains of a thousand documents each randomly selected from the Reuters21578 corpus. The experimental
results demonstrate improved coherence of the clusters produced by SETS
compared to the text normalization obtained with the Porter stemmer. In addition, semantic-based text normalization is shown to be resistant to noise, which
is often introduced in the index aggregation stage.

Keywords: Semantics, stemming, cluster coherency, partitional clustering.

1 Introduction

Document clustering is employed in the domain of information retrieval (IR), e.g.
search engines which return a set of similar documents that resemble a query of words
to a certain extent [1]. Studies of this domain indicate a growing need for more efficient clustering, to facilitate document browsing and knowledge discovery [2, 3].
Document clustering relies on features acquired from text to measure the pair-wise
document similarity. The features called word stems are obtained from documents
after text normalization.

Normalization is traditionally conducted by the Porter stemmer [4], which employs
suffix stemming [5]. It involves rule-based transformations, which remove known
suffixes of words by relying on language morphology. The words are stemmed to
their morphological root form. Rule-based normalization recognizes as similar those
words that share a common grammatical root but it may produce errors as a result of
over- or under-stemming [6]. The Porter stemmer provides a good trade-off between
speed, reliability and accuracy. State-of-the-art algorithms, which perform slightly
better and provide little advantage over the Porter stemmer, are slower and more

M. Graña et al. (Eds.): KES 2012, LNAI 7828, pp. 108–118, 2013.
© Springer-Verlag Berlin Heidelberg 2013

difficult to implement [7]. The stemming efficiency depends on the computational complexity and the quality of the corpus.

This paper focuses on increasing the coherency of the clusters and providing better groupings by developing a semantic text normalization technique, and using semantic hierarchies in the text representation to provide more abstract indexing. The semantic stemmer proposed in this paper is based on the Porter stemmer, which is used as a baseline stemmer in the evaluation.

The rest of the paper is organized as follows. Section 2 discusses related work in the domain of clustering and reviews traditional document clustering. It also introduces the spherical k-means clustering, which is used in the evaluation. The proposed semantic clustering algorithm is discussed in Section 3; its evaluation is presented in Section 4. Conclusions with regard to the performance of the proposed algorithm are outlined in Section 5.

2   Related Work

This section firstly discusses feature extraction of word stems from text documents using the Porter stemmer. Then the traditional vector space model representation is analyzed. Finally, the standard k-means algorithm with cosine similarity measure, i.e. spherical k-means, used in the evaluation, is described.

2.1   Feature Selection and Feature Extraction

Partitional algorithms use words as features selected for document representation. The features used in the representation are the word stems; they are aggregated by applying series of transformations. The transformations involve pre-processing of the documents by employing stemming algorithms, such as the Porter stemmer. The text normalization is followed by calculating the statistical co-occurrence of the words. Then the index aggregated for the document representation is created. It is represented as a set of pairs for each document. Each pair consists of a word stem and its weight, $< s, w >$. Feature selection and extraction are crucial to effective and efficient clustering. Good features can result in decreased workload and simplified subsequent clustering or/and improved document groupings [8].

Porter Stemmer
The Porter stemming algorithm is the word normalization technique most widely used by the information retrieval community. It provides normalization at document level by removing the common morphological and inflectional endings of the words. The stemmer produces a reduced number of root elements (it reduces dimensionality) by conflating a group of words into a single root element. A group of conflated words is believed to indicate the same topic. Although the morphological forms of the produced words are not necessarily real words, the documents retrieved indicate good retrieval quality [9].

The Porter stemmer employs various suffix-stripping rules and does not rely on an external lexical source, which affects the accuracy of the stemming. Additionally, the algorithm does not handle well words that do not follow the morphological rules of word inflection, i.e. irregular verbs and words that shift from their root form when a suffix is added. In the case of the words "wand" and "wander" as well as "experience" and "experiment", the algorithm wrongly conflates the words respectively to "wand" and "experi". The ending "-er" of the word "wander" is considered a suffix and is stripped off, which changes the meaning of the word. Instead, the algorithm should leave the ending untouched and consider the whole word as part of the stem. In the case of "experience" and "experiment" the change in the meaning of the words occurs as a result of ambiguity rather than wrong meaning. It is difficult to construct rules that cope with a rare change of the root form of the word when a suffix is added.

These shortcomings illustrate the need to improve the Porter stemmer. Stripping the suffixes of the words without using word sense disambiguation and/or part of speech disambiguation, introduces ambiguity. In addition, the stemmer needs to handle or avoid the rare inflected irregular forms of the words produced using conjunctions. Context sensitive stemming [10] is used for document search, where the corpus is analyzed to find the distributional similarity of the words. The next step is to apply the Porter stemmer on candidates acquired from the text to remove the grammatical inflections of pluralization. The forms obtained from the words are used in query expansion on non-transformed index to retrieve documents. Derivational and inflectional stemmers improve the Porter's algorithm by adding a dictionary check after each iteration [11]. They stop the stemming if a correct form of a word is found. Thus, these stemmers are able to process irregular forms. The resulting stemmer however, performs worse than the original Porter stemmer at additional computational cost.

2.2   Document Representation

Partitional clustering employs the vector space model (VSM) that treats documents as a bag of words (BOW) [12]. The documents in a collection are transformed into VSM using TF-IDF weights. The weight of stems that do not occur in the document is 0. This transformation into VSM yields a matrix, where each row is the vector representation of each document from the corpus in the TF-IDF vector space. The dimensionality of the matrix is usually very high and makes the scalability of the clustering algorithms difficult [13]. The scalability problem is typical for algorithms that produce good results on a small dataset or in a specific domain but fail to perform on larger scale or across domains. The latter is due to the ambiguity of words and the fact that many domains share common terms, which may contribute to low quality in the document groupings [14].

A strategy for achieving better performance and scalability of the clustering is achieved by combining traditional feature selection with a semantic-based approach to dimensionality reduction. The semantic approach, which relies on a general ontology, is employed in large scale indexing of web pages with concepts. It uses a higher order semantic hierarchy in the document representation vectors and is regarded as concept indexing [15]. It considers all possible meanings of the words. A word with multiple meanings shares its weight (significance) equally among the concepts it belongs to. In the end, an accumulated scoring result for every possible concept is calculated (see formula 1).

$$w_c(d_j) = \sum_{i=1}^{n} \left(w_{tf-idf}(t_i, d_i) \frac{1}{C_{(t_i)}} \right) \tag{1}$$

where n is the number of terms in the document that contains a concept C, $w_{tf\text{-}idf}$ denotes the significance of a stem. The coefficient $1/C_{(t_i)}$ represents the "empirical observations and the idea that monosemic words are more domain-oriented than polysemic ones and provide a greater amount of domain information" [15]. The index aggregated for the document representation comprises pairs of concepts and weights, $< c, w >$ for every document. The size of the vectors does not exceed 990, which is the number of concepts in the ontology used (see section 3.1).

Similarly, a method that employs an ontology to "enrich the term vector with concepts" is proposed by Setchi et al. [15]. The use of higher order structure to replace the words in the representative vectors with concepts significantly reduces dimensionality and computational complexity. In addition, the higher order hierarchy provides better scalability and improved clustering. This approach provides a generic perspective in establishing similarity between topics when used in clustering [15]. The clusters produced are expected to have better coherency.

2.3   Standard Document Clustering with k-Means

In order to apply the k-means algorithm, the document collection is represented in the VSM, where each document is a vector d in the vocabulary space. The position of the vectors in the multidimensional space is defined by the co-occurrence of the terms from the collection within the documents, i.e. TF–term frequency, multiplied by the inverse document frequency (IDF) of the terms. Thus, TF-IDF defines the representative weight of each term within each document in the collection:

$$d_{tf-idf} = (tf - idf_1, tf - idf_2, tf - idf_3, ..., tf - idf_n) \tag{2}$$

where $tf - idf_i$ is the weight of the token with index i. As a result the more frequent words in the collection have less discriminative power (less IDF weight) than the more infrequent words. A word that is a representative token of the collection but is not in a specific document has a weight of 0. However, computing the weight of all words across all documents leads to high computational complexity.

Partitional clustering creates clusters with flat, non-hierarchical structure. The number of clusters is controlled by a parameter passed to the algorithms prior to execution. This parameter k drives the process of partitioning documents in k clusters by employing the k-means algorithm. Selecting the number of clusters without a priori domain knowledge in the area of interest may worsen the results. In addition, if documents cover a broader thematic area, the clusters can be inferior [16].

The spherical k-means algorithm uses the robust cosine measure to compute the similarity between documents. It is defined as

$$cosine(d_1, d_2) = (d_1 \cdot d_1)/\|d_1\| \, \|d_2\| \qquad (3)$$

where $\cdot$ denotes the vector dot product and $\|d\|$ is the length of the vector. The k-means algorithm randomly picks k centroid vectors to identify the closest documents to the centroids and forms clusters around them. The algorithm iteratively refines the randomly chosen initial k centroids, minimizing the average distance between them. The algorithm is not deterministic.

3   Semantically Enhanced Text Stemmer (SETS)

In this section first, the hierarchical structure of the ontology used in this research (OntoRo) is presented. Then, the semantically enhanced feature extraction algorithm is introduced. Finally, a document representation with reduced dimensionality is discussed.

3.1   OntoRo

The specific task of the proposed algorithm, to group documents on various topics in coherent clusters using semantics, requires the use of a general ontology. The general ontology provides a generic perspective on the relations between the concepts in the documents. These relations are used to estimate the distance between pairs of documents. The assumption of the algorithm is that by using the generic relations established between the words in the selected ontology, similar relations between the documents could be detected, which will result in improved cluster coherence.

The general ontology OntoRo resembles the structure of a thesaurus. The words are grouped together based on their meanings in a hierarchy of concepts, heads, sections and classes [17]. This makes OntoRo's structural organization very rigid and robust in view of the fact that it is tree based. Every class in OntoRo is a root of a separate tree. It is important to note that the word groupings provided by OntoRo are based on the ideas these words convey and not on synonymy-based relations. There is no specific conceptual interconnectivity that pre-defines these relations.

The proposed algorithm employs concept indexing as a text representation technique and OntoRo as an external lexical source. For the purpose of the stemming in the next section, OntoRo is stemmed using the Porter stemmer. Thus, all words in OntoRo are stemmed to produce a second lexical source regarded as stemmed Onto-Ro. An example for the word *connect* is shown in Table 1. The first column contains all inflectional forms of *connect* which can be found in OntoRo. The column 'Root Form' contains the relevant root forms of the words from column 'Word' produced by the Porter stemmer. The column 'Semantic Meanings' shows the semantic polysemy of the words represented with OntoRo. Since the words in column 'Word' conflate to the morphological root *connect*, the stemmed OntoRo will contain only the word ‚connect‘, which has 12 unique meanings presented in table 1 for all forms of *connect[1]*. For that reason, the semantic stemming proposed in the next section aims to conflate the words to less ambiguous morphological forms.

Table 1. Semantic representation of word stems in OntoRo

Word	Root Form	Semantic Representation[2]	Semantic meanings
connect	connect	$C_9, C_{45}, C_{62}, C_{71}, C_{202}, C_{305}$	6
connected	connect	C_9, C_{45}, C_{50}	3
connection	connect	$C_9, C_{11}, C_{45(2)}, C_{47}, C_{48}, C_{706}$	7
connecting	connect	C_{45}	1
connective	connect	C_{45}, C_{47}	2

3.2   Semantic Stemming Algorithm

In this section an algorithm is proposed, which relies on semantics to achieve text consistency. It is called a semantically enhanced text stemmer (SETS). A semantic approach is used to address the problem of words, which are conflated to the same morphological stem but have different semantic representations. Table 1 shows inflectional forms of the word *connect* with different semantic representations in OntoRo, which refer to different meanings. However, they are stemmed to the same morphological root by the Porter stemmer. This makes the stems more polysemic and when used in clustering the clusters produced are inferior. The proposed algorithm aims to improve cluster coherency by keeping the normalized words as monosemic as possible.

The algorithm listed in Fig 1 extends the six steps of the Porter stemmer [4]. Every word that occurs in a document is searched first in OntoRo by its morphological form. If the word is found it is considered stemmed and the algorithm proceeds to the next word. If the word does not occur in the lexical source, the algorithm proceeds with the first step of the algorithm in Fig 1. The stem produced by the first step of the algorithm is searched for occurrence in OntoRo. If the stem is found it is considered to be stemmed. This process is repeated for each of the six steps of the algorithm in Fig. 1. In case a word is not stemmed after the last 6th step it is searched for occurrence in the stemmed OntoRo. Finally, if the algorithm does not find the stem in the stemmed lexical source, it is considered a named entity and no semantic information is acquired. Otherwise, the algorithm returns all concepts that the stem is associated with. Every concepts refers to a different meaning of the word.

---

[1]   *connect* in stemmed OntoRo $C_9, C_{11}, C_{45(2)}, C_{47}, C_{48}, C_{50}, C_{62}, C_{71}, C_{202}, C_{305}, C_{706}$.

[2]   The semantic representation $C_{<number1>(<number2>)}$ stands for concept (C), concept number in OntoRo ($< number1 >$) and the number of semantic meanings in the concept ($< number2 >$) (in case of one semantic meaning the number is omitted), i.e. $C_9 - relation, C_{11} - consanguinity, C_{45} - union, C_{47} - bond, C_{48} - coherence,$ $C_{50} - combination,$ $C_{62} - arrangement, C_{71} - continuity, C_{202} - contiguity, C_{305} - passage, C_{706} - cooperation;$

The next stage is to use the semantic stems produced by the SETS algorithm to represent the documents using VSM. For this purpose, the weights (TF-IDF) for all stems in the collection are calculopated for every document. Then, using equation (1) the documents are represented in the higher order hierarchy of concepts yielding a matrix of concept indices $< concept, weight >$. The concept indexing reduces dimensionality, since the concept number is limited to the number of concepts in OntoRo.

```
in:    w ... word
out:   s ... set of semantic meanings
s = ontoro_search_for_occurrence(w)
for(step = 1; s is {} and step <= 6; step = step + 1)
   w = porter(w, step)
   s = ontoro_search_for_occurrence (w)
end
if s is {}
   s = ontoro_stemmed_search_for_occurrence (w)
end
# if s is {}, w is a named entity
# otherwise s contains the semantic meanings of W
return s
```

Fig. 1. Semantically enhanced text stemming algorithm

This approach is similar to using a dictionary for searching the word stem after every step of the Porter stemmer [11]. However, the aim of the proposed algorithm is to acquire less ambiguous semantic information and not real words. This is achieved by considering the semantic meaning of the words before applying rule-based stemming. The errors from under- and over-stemming are thus alleviated. The proposed algorithm still relies on TF-IDF to measure the discriminative power of words.

4   Evaluation and Discussion

The evaluation of the proposed SETS algorithm is performed using the Reuters-21578 text categorization test collection. It consists of 21578 news articles on different economic subjects published in 1987. The focus of this paper is to investigate the enhancement that semantics provides in text normalization, and document clustering in particular. For the purpose of the evaluation, the corpus is transformed into VSM twice, first by employing the proposed SETS algorithm and then by employing the classical approach of TF-IDF weighting after the Porter stemmer normalizes text. The latter transformation produces tokens that represent normalized word forms, i.e. Porter stems. The implementation used in the evaluation is the Common Lisp version of the algorithm made available by Porter. This transformation into VSM yields two matrices where each row is the vector representation of a Reuter's news article in word-based (TF-IDF weights) or respectively OntoRo vector spaces. The TF-IDF matrix is a sparse matrix with 21578 rows and 44293 columns – one for each unique word in the corpus. The SETS matrix has 21578 rows and 990 columns; one for each concept in OntoRo and this matrix is dense. The collection was reduced to 18457 documents due to the realization that 3121 documents (21578–18457) are either very short and/or contain only named entities and cannot benefit from concept indexing. The document-term and –concept matrices are produced for the 18578 documents. Then, a cross-domain clustering is simulated by constructing 20 sub-collections of a thousand files (rows) each. By design, the construction of a sub-collection introduces noise in the document representation, since the statistical analysis (calculating the IDF values for the words) is carried out on all 18578 documents and not on the subset of a thousand documents in the sub-collections. Thus, the documents representations used in the clustering process are tested for robustness.

All matrices are clustered using the spherical k-means algorithm [33], which is available in the CRAN repository. This version of the algorithm is fast and requires the number of clusters as input. Experiments are performed to split the data in 5, 10, 15, 20, 25, 30, 35, 40, 45, and 50 clusters. The quality of the clusters is assessed with the silhouette measure proposed by Rousseeuw [3]. This measure is defined as

$$s(i) = \frac{b(i)-a(i)}{max\{a(i),b(i)\}} \tag{4}$$

where $a(i)$ is the average dissimilarity of the object (row) i compared to all other objects of a cluster A and $b(i)$ is the average dissimilarity of the object i element A compared to all objects in the cluster nearest to i. To assess the overall quality of a clustering model (M), the measure $p(M)$ was averaged over all objects:

$$p(M) = \frac{1}{n}\sum_1^n s(i) \tag{5}$$

where n is the total number of news articles. The measure is a number between 0 and 1 with higher number suggesting stronger cluster coherence. Table 2 shows the statistical average values for all sets including the standard deviation. The statistical analysis of the obtained results is needed since the sk-means algorithm is not deterministic. The results demonstrate that the SETS algorithm is outperformed by the TF-IDF weighting approach and Porter when the number of clusters is small – the reduced dimensionality of document representation provided by the SETS cannot separate clusters well for a small number of clusters when there is a large number of files. This is due to the dense document-concept matrix, i.e. most of the concepts are used to represent the documents. As the number of clusters increases, the performance of the SETS algorithm as well as the separation of the clusters improves. In addition, the hypothesis that semantic stemming of the words, i.e. semantic-based word disambiguation, leads to clustering solutions with better coherence is proven right.

Table 2. Statistical Mean and Standard Deviation for all clustering solutions

Clusters N	Porter		SETS	
	Mean	SD	Mean	SD
5	0.0658	0.0134	0.0299	0.0036
10	0.0663	0.0112	0.0446	0.0041
15	0.0643	0.0133	0.0547	0.0054
20	0.0662	0.0129	0.0624	0.0058
25	0.0643	0.0123	0.0706	0.0104
30	0.0587	0.0162	0.0766	0.0066
35	0.0610	0.0113	0.0821	0.0080
40	0.0526	0.0120	0.0866	0.0070
45	0.0510	0.0086	0.0922	0.0049
50	0.0466	0.0111	0.0968	0.0068

The clustering solutions produced from both matrices use the natural dimensionality of the document representation, i.e. no dimensionality reduction techniques are used. The SETS represents documents in reduced dimensionality by design. However, the execution times shown in Table. 3 indicate slower clustering of the SETS data. S-k means produces faster clustering from the sparse TF-IDF matrix where 99.99% of it is populated with zero values. In addition, to produce the dense document-concept matrix the SETS algorithm requires 5.11 times more time than the Porter stemmer. This is as a consequence of searching the stems in OntoRo after each step of the Porter's algorithm. The memory footprint of the entire SETS matrix is 7.02MB, whilst TF-IDF in dense representation is 1.7GB and 25MB in sparse representation.

Table 3. Execution time in seconds for the clustering solutions

	5	10	15	20	25	30	50
TF-IDF	8.15	27.76	28.60	34.01	47.78	53.66	85.64
SETS	34.95	54.19	99.42	43.15	121.36	103.42	104.24

5   Conclusions and Future Work

This paper presents a comparison of the Porter and the SETS stemmers when employed by the partitional document clustering algorithm sk-means in cross-domain environment. The SETS algorithm represents a semantically enhanced Porter stemmer where the latter provides good trade-off between speed, reliability and accuracy. The other text stemmers either achieve worse stemming results, i.e. higher dimensionality of document representation, or better results at higher computational cost, i.e. they are not suitable for large collections. SETS aggregates semantic information from text and uses it to represent documents in more abstract and generic manner. According to the results, the sk-means clustering algorithm achieves better and more consistent cluster coherence regardless of the choice for the parameter k. In addition, the obtained document representation produced by the SETS algorithm is proven to be more resistant to noise, which is usually introduced in the index aggregation phase, than the standard document representation provided by the Porter normalization and the TF-IDF weighting mechanism.

A limitation of the algorithm is that it is designed to work well on generic cross-domain document collections. In collections, where documents are topically grouped based on named entities, the algorithm is expected to perform worse. However, the performance can be improved by using a domain specific ontology, which provides specific semantic information referring to the specific terms of the domain of interest or named entities. Alternatively, word-based clustering algorithms would be a more adequate tool. In this paper, the choice of the parameter k was made with the purpose to explore a broad range of the parameter space without considering the optimal number of clusters for the Reuters21578 data.

The evaluation demonstrates that SETS in conjunction with concept indexing provides reduced dimensionality of document representation. This enables more advanced clustering algorithms to be used to produce clustering solutions on a large scale. In addition, the approach can employ a domain specific ontology along with OntoRo to produce clustering solutions from the perspective of the used domain knowledge. The scalability and the flexibility of this approach in terms of number of documents and different perspectives to a collection of documents provides opportunities for developing new approaches to grouping documents together and discovering relationships between them.

References

1. Jain, A.K., Murty, M.N., Flynn, P.J.: Data Clustering: A Review. ACM Computing Surveys 31(3), 264–323 (1999)
2. Cutting, D.R., et al.: Scatter/Gather: a cluster-based approach to browsing large document collections. In: Proceedings of the 15th Annual International ACM SIGIR Conference on Research and Development in Information Retrieval 1992, Copenhagen, Denmark (1992)
3. Carpineto, C., et al.: A survey of Web clustering engines. ACM Computing Surveys 41(3), 1–38 (2009)
4. Porter, M.F.: An algorithm for suffix stripping. In: Jones, K.S., Willett, P. (eds.) Readings in Information Retrieval, pp. 313–316. Morgan Kaufmann Publishers Inc., San Francisco (1997)
5. Jongejan, B., Dalianis, H.: Automatic training of lemmatization rules that handle morphological changes in pre-, in- and suffixes alike. In: Proceedings of the Joint Conference of the 47th Annual Meeting of the ACL and the 4th International Joint Conference on Natural Language Processing of the AFNLP, vol. 1. Association for Computational Linguistics, Suntec (2009)
6. Xu, J., Croft, W.B.: Corpus-based stemming using cooccurrence of word variants. ACM Trans. Inf. Syst. 16(1), 61–81 (1998)
7. Smirnov, I.: Overview of Stemming Algorithms. Mechanical Translation (2008)
8. Xu, R., Wunsch, D.: Survey of clustering algorithms. IEEE Transactions on Neural Networks 16(3), 645–678 (2005)
9. Wessel, K., Ren, P., et al.: Viewing stemming as recall enhancement. In: Proceedings of the 19th Annual International ACM SIGIR Conference on Research and Development in Information Retrieval 1996. ACM, Zurich (1996)

10. Lee, L.: Measures of distributional similarity. In: Proceedings of the 37th Annual Meeting of the Association for Computational Linguistics on Computational Linguistics 1999. Association for Computational Linguistics, College Park (1999)
11. Krovetz, R.: Viewing morphology as an inference process. In: Proceedings of the 16th Annual International ACM SIGIR Conference on Research and Development in Information Retrieval 1993, Pittsburgh, Pennsylvania, United States (1993)
12. Salton, G., McGill, M.J.: Introduction to Modern Information Retrieval. McGraw-Hill, Inc., New York (1986)
13. Fung, B., Wang, K., Ester, M.: Hierarchical document clustering. In: The Encyclopedia of Data Warehousing and Mining. Idea Group, NY (2005)
14. Gliozzo, A., Strapparava, C., Dagan, I.: Unsupervised and Supervised Exploitation of Semantic Domains in Lexical Disambiguation. Computer Speech and Language 18(3), 24 (2004)
15. Setchi, R., Tang, Q., Stankov, I.: Semantic-based information retrieval in support of concept design. Advanced Engineering Informatics 25(2), 131–146 (2011)
16. Steinbach, M., Karypis, G., Kumar, V.: A comparison of document clustering techniques. In: KDD Workshop on Text Mining, Boston (2000)
17. Setchi, R., Tang, Q., Bouchard, C.: Ontology-Based Concept Indexing of Images. In: Velásquez, J.D., Ríos, S.A., Howlett, R.J., Jain, L.C. (eds.) KES 2009, Part I. LNCS, vol. 5711, pp. 293–300. Springer, Heidelberg (2009)

Ontology Recomposition

Tomasz Rybicki

University of Warsaw, Poland
rybicki.uw@gmail.com

Abstract. Electronic devices provide many different kinds of simple services. Most of them is able to interoperate with each other. In order to provide sufficient quality of service discovery and composition, ontologies that are used to describe services should contain many concepts and relations, which makes them big and thus unsuitable to be processed mobile devices. To fully exploit the power of a service-oriented approach in a pervasive computing environment, a technique of runtime ontology recomposition is proposed. Service descriptions created from a common ontology are composed into a runtime version of the ontology. This runtime ontology is then used to resolve service queries instead of the original one. The descriptions' semantic richness influences the quality of service discovery and composition. In order to assess their most favourable richness the original ontology is used.

Keywords: semantic services, ontologies, ontology engineering, ontology recomposition.

1   Introduction

In our daily environment we are surrounded by electronic devices. They come in different shapes and sizes, but, in essence, all of them are small computers that can provide simple services: audio playback, GPS-based localization, picture & video capture...etc. Those services may be invoked individually, but they may also be composed into complex services, producing richer user experience. Such an interoperability is not new – semantic web services (SWS) [1] are ontologically annotated web services (WS) [2] that may be automatically discovered, composed and executed. SWS, however, run on powerful web servers and are used to embody complex business logic. Pervasive computing environments, on the other hand, are constituted of mobile nodes that provide services used and composed on-the. Using ontologies to compose services in such environment faces some challenges – to provide a good quality of service discovery and composition, an ontology should contain a lot of information about relationships between services. This translates into its size, which in turn, may impede its processing in devices with limited computing capabilities. Therefore, some way of reducing the size of ontological representation of services is required.

   An ontological representation of a domain of interest consists of two components: a TBox and an ABox. The TBox describes the domain in terms of concepts and

M. Graña et al. (Eds.): KES 2012, LNAI 7828, pp. 119–132, 2013.
© Springer-Verlag Berlin Heidelberg 2013

relations, while the ABox in terms of instances of those concepts. Taking the service-oriented approach, the TBox contains definitions of service types, while the ABox descriptions of actual service implementations.

A straightforward approach to describe a service environment at runtime is to store full ontological representation, i.e. information about all the types and all the instances, on each node (Fig. 1a). Such a node description would consist of a full TBox, extended with full ABox containing information about all deployments of all services present in the environment. This technique is unrealistic, if only due to network reconfiguration problems. Another scenario is to store on a node only those ABox assertions that are related to this specific node (Fig. 1b): such a description would consist of a full TBox extended with ABox assertions defining services available in the node. This solution, still has a severe drawback – full TBox may consist of hundreds of concepts and its size may still be significant. In this work, full ontological representation of an environment is decomposed into individual node descriptions (so called TBox snippets [3]). Such a description consists of concepts directly related to the service being described and of some additional concepts, called 'enrichment'.

Descriptions of nodes present in the environment are composed at runtime into a runtime version of the ontological representation of the environment. This runtime ontology thus consists of descriptions of relevant services only. In order to provide that service discovery and composition performed against runtime ontology will deliver similar results as performed against full ontology, a sufficient level of 'enrichments' (which might be easily assessed) in service descriptions must be maintained.

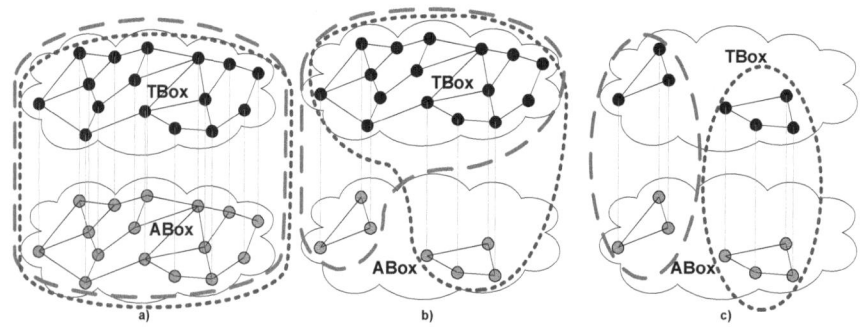

Fig. 1. Outlines (dotted and dashed) represent node descriptions. A straightforward (a) approach is to embed full ontology in each node. A more realistic scenario (b) is to reduce the ABox to assertions concerning the node (c). In the approach described in this paper both the TBox and ABox parts embedded in a node contain only statements concerning the node.

The paper structure is as follows: section 2 contains description of the related work in the area of ontology recomposition and service matchmaking. In sections 3 and 4 structure of the ontological representation of services and service descriptions are explained. The ontology recomposition technique and the discussion of its impact on the quality of service discovery and composition is presented in section 5. Section 6 concludes the paper.

2   Related Work

2.1   Service Matchmaking

Service matchmaking is a basis for service discovery a composition – a query is matched against service descriptions in order to find services, either individual or complex that satisfy it. The first (chronologically) and the simplest techniques are text-based approaches [4-7]. Their primary use is to facilitate human-based service matching. Service descriptions take the form of natural language definitions. Services are annotated with labels and matchmaking is performed by employing string matching. This reduces its precision (i.e. it is possible to retrieve services whose descriptions contains some but not all of the query keywords). Also, the lack of semantic understanding results in reduced recall as some services that are semantically compatible with the query might be rendered not matching it (e.g. services described with synonyms), while services semantically not compatible (e.g. services described with homonyms) might be rendered matching.

Some of the problems described above might be resolved by employing a logic language in service description [8-15]. By providing formal descriptions, the matchmaking becomes unequivocal. On the other hand, such approaches take the closed-world assumption, meaning that they need a predefined representation of states, actions, objects passed, etc. Each state, action that is not predefined is assumed false or invalid. This prevents acquiring new types of information during system lifetime and thus limits service opportunities to only the foreseen ones. Also, although the representation of entities might be reasonably simple, they still require complex data and control-flow structures (e.g. if-then) to express composite services, which makes them resource-consuming.

Employing ontologies in service matchmaking [8, 12, 16-23] overcomes some of the problems described above. By exploiting relations between concepts describing services, a relationships between services might be detected which adds to flexibility of service matchmaking (thus allowing serendipitous service discovery) without sacrificing its quality [24]. Unfortunately, recent research [25] indicates that, similarly to logic-based approaches, the semantic matching of service descriptions is a heavy process.

The approach presented in this work employs semantic matchmaking. It shares the most features with those in [8, 20]; however, only inputs and outputs are considered and no match degrees are used: outputs in the request must subsume or be equivalent to concepts in the advertisement and inputs in the request must be subsumed or be equivalent to concepts in the advertisement.

2.2   Ontology Decomposition

The goal of ontology decomposition is to obtain one or more subsets of the original ontology. Existing approaches may be divided into three categories: query-based methods, segmentation techniques and traversal-based extractions.

Query-based methods [26-30] provide an SQL-like interface to ontology. They offer a simple, low-level access to the semantics of the ontology. The produced extract is transient, as it usually (e.g. in [29]) may not be updated, shared or modified independently from the ontology it was taken from. Also, its further integration into a new ontology is difficult.

Segmentation techniques [31,32] divide ontology into many independent modules. The modules might overlap, but they are either used separately or only loosely coupled. The algorithms are lossy. While they might ensure the completeness of reasoning in a context of a module, by discarding less relevant concepts or properties, some of the semantics of the whole ontology is irrevocably lost as the original ontology is destroyed during the process. The goal is to decompose the ontology into a number of inter-related islands [33] of concepts. The techniques divide ontology based on abstract attributes, e.g. desired number of modules. Ontology segmentation techniques are best suited for automatic processing of ontologies of different sizes and structures.

The ontology-traversal extracts, similarly to query-based approaches, are centered around a predefined set of concepts and/or relations. The difference, however, lies in the fact that they are application-specific. In [34], for instance, only subsumption and OWL restrictions relationships are considered to link concepts in the ontology. This way, the generality of the approach is reduced. On the other hand, they allow to further fine tune the extracted sub-ontology depending on the specific application. This is the case of [35] where certain concepts/relations are omitted in order to produce more concise and compact definitions of provided terms.

While query-based methods are too specific for the task of ontology recomposition, the query segmentation techniques are too generic. Although their behavior might be fine-tuned, the control is still too coarse. The sub-ontology extraction method which is a part of ontology recomposition technique presented in this work follows the traversal-extraction methodology. It is targeted at a specific usage - creation of individual service descriptions from a large ontology. This, along with the fact the base ontology has a predefined structure imposes specific characteristics and structure on the extracts, which facilitate their later integration into a new ontology.

2.3   Ontology Composition

Existing approaches to ontology composition might be divided into three types – ontology integration, ontology merging and ontology alignment [36-38]. Ontology integration and ontology merging are processes of generating a single, coherent ontology from different (sub)ontologies: ontology integration deals with ontologies covering different subjects while ontology merging deals with ontologies covering the same subject. Ontology alignment is the process of creating links between different ontologies. They all rely on ontology mapping – a process of relating identical concepts or relations in different ontologies.

There are a number of approaches of finding mappings between concepts and relations in different ontologies. The most notable group of approaches relies on text manipulation techniques [39-42]. The similarity between concepts is calculated based on similarity in their naming, either directly (the names are equal) or with the use of text-processing techniques like keyword extraction (e.g. looking for certain nouns in concepts descriptions), language processing (e.g. looking for different declanations of certain nouns) or the use of linguistic resources (e.g. the use of lexicons or thesauri to identify synonyms).

Another way of detecting similarity between concepts from different ontologies is to compare their definitions [39, 40, 43]. This amounts to comparing their attributes, properties, instances or related classes. Also, structure of the ontologies being compared might be analysed [39, 40, 41]. Such approaches treats concepts as vertexes and properties as edges in graphs and apply graph-based algorithms to detect similarity.

In this work, (sub)ontologies that were previously extracted from the same or similar (in terms of its structure) ontology are merged. The (sub)ontologies are then merged into a runtime version of the starting ontology. Such runtime ontologies contain less concepts and properties but cover the same subject (service description) and have similar structure. Concepts with the same names are considered equal.

3   A Common Ontology

An ontological representation of a selected domain (a common ontology) consists of two layers [3]: an upper level ontology that defines a service model and a domain ontology. The upper ontology contains 'marker' concepts that denote atomic services (`AtomicService`) and their parameters (`ServiceParameter`). The domain ontology subclasses of the former represent actual service types (e.g., `FireDetector`), while the domain ontology subclasses of the latter represent data types used as service parameters (e.g., `Temperature`). An example of a common ontology is shown in Fig. 2. Rounded rectangles denote ontological concepts (service and parameter types), arrows denote concept subsumption (`are`) or link service and parameter type concepts (`hasInput`, `hasOutput`). Such ontology consists of three parts: (1) a service type hierarchy used by the user to organize services (shown in Fig. 2a), (2) a parameter type hierarchy used to discover services and compose them into complex ones (Fig. 2b), and (3) axioms linking both hierarchies (Fig. 2c). Relations between service and its parameter types are expressed by `hasInput` or `hasOutput` properties: the former denotes an input parameter, the latter denotes an output parameter.

Common ontology is processed to find services matching input-output criteria defined in a query. The matchmaking algorithm [44] is based on those proposed in [10] and [20]. It starts with identifying services whose outputs match those in the query: a service's output parameter type must be subsumed by the output type in the query. Then, those services are inspected with respect to their inputs: a service's input parameter type must subsume an input type in the query. Also, a desired service must not require any other input types other than those in the query. The algorithm allows to compose services either vertically (outputs of one service become inputs to another service) or horizontally (a set of services is gathered that together provide all required outputs).

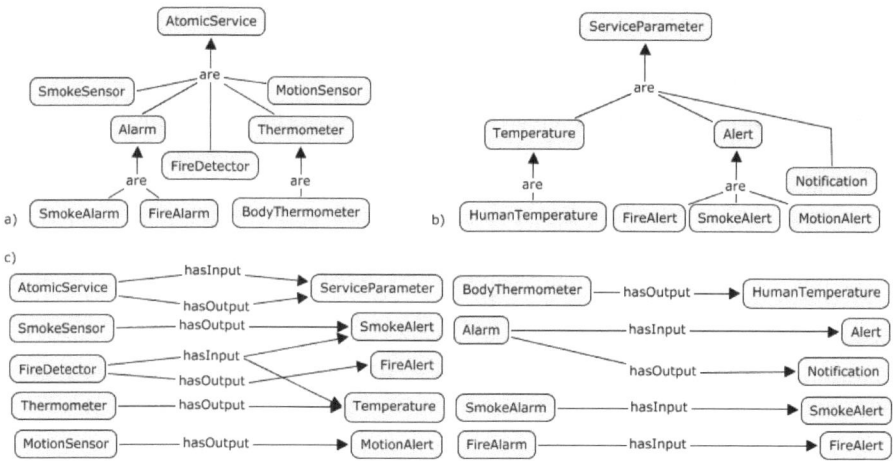

Fig. 2. An example of a domain ontology. The hierarchy of service types (a), the hierarchy of parameter types (b) and relations linking the hierarchies (c).

4   Service Description

Each device is characterized by the type of the service it provides – its description is created from a common ontology and consists at least of one concept that represents a service type and one or more concepts representing parameter types[1]. Fig. 3 shows examples of descriptions for the Thermometer (Fig. 3a) and FireDetector (Fig. 3b) services.

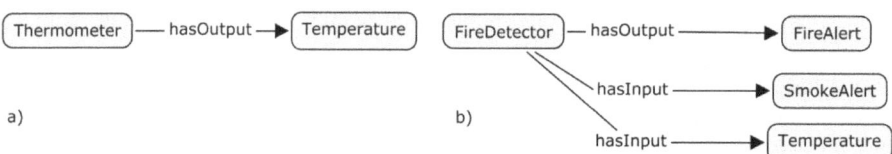

Fig. 3. Service descriptions (core): (a) Thermometer and (b) FireDetector

A description might consist of concepts representing service and its parameter types only ('core' description), but it may also contain some additional concepts (an 'enrichment'). Level 1 enrichment consists of concepts being parents and children of concepts in the core description, level 2 enrichment of concepts being grandparents and grandchildren.....etc . In this context, service descriptions in Fig. 3 have level 0 enrichment (in short: richness 0). Fig. 4 shows the same descriptions with level 1 enrichment (children and parents of the service and service parameter concepts are included in a description).

---

[1] Term 'parameter' is used to refer to both inputs and outputs of a service.

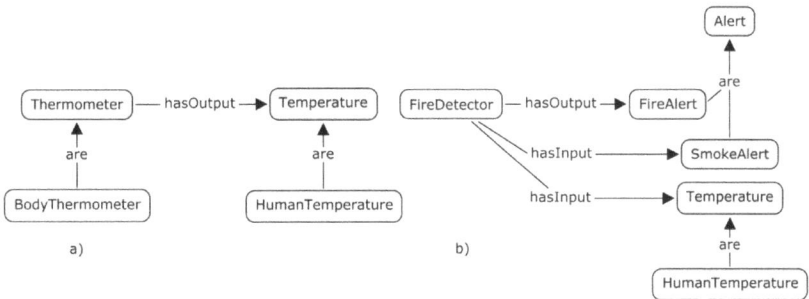

Fig. 4. Service descriptions (richness 1): (a) Thermometer and (b) FireDetector

While devices with more computing power may host 'richer' descriptions, simpler devices will be provided with 'core' descriptions only. It is also important to adjust the enrichment level to the intended environment where the service will be used: while the additional concepts improve the quality of service discovery and composition, above certain enrichment level they only inflate the size of ontological representation and thus increase the 'cost' (in terms of the required computing power) of processing the ontology.

5   Recomposing an Ontology

The goal of the ontology recomposition technique is to create runtime version of a common ontology that contain concepts and relations closely representing the state of an environment. This runtime ontology (RO) is composed on-the-fly from individual service descriptions, embedded in devices present in the environment [3]. RO is thus fine tuned to the environment as it does not contain description of services that are unavailable (e.g. due to the absence of the device-hosting them). Also, the enrichment level of the constituent service descriptions is transferred to the RO and impacts its service matchmaking capabilities.

5.1   The Importance of the Enrichment Level

Runtime ontology is composed from individual service descriptions present in a specific environment. If all service descriptions have no enrichments, there is a probability that when composed into RO, some of them will not share concepts or object properties with other descriptions (Fig. 5a). Since the matchmaking algorithm employs relations between concepts, this may limit service's composability – such isolated descriptions cannot become a part of a complex service and may only be used alone. With the addition of enrichments, the likelihood of sharing concepts/relations with other descriptions increases (Fig. 5b). The more of them the better the service discovery and composition. It is thus essential for the service descriptions to 'overlap', i.e. to contain as high enrichment level as possible (Fig. 5c).

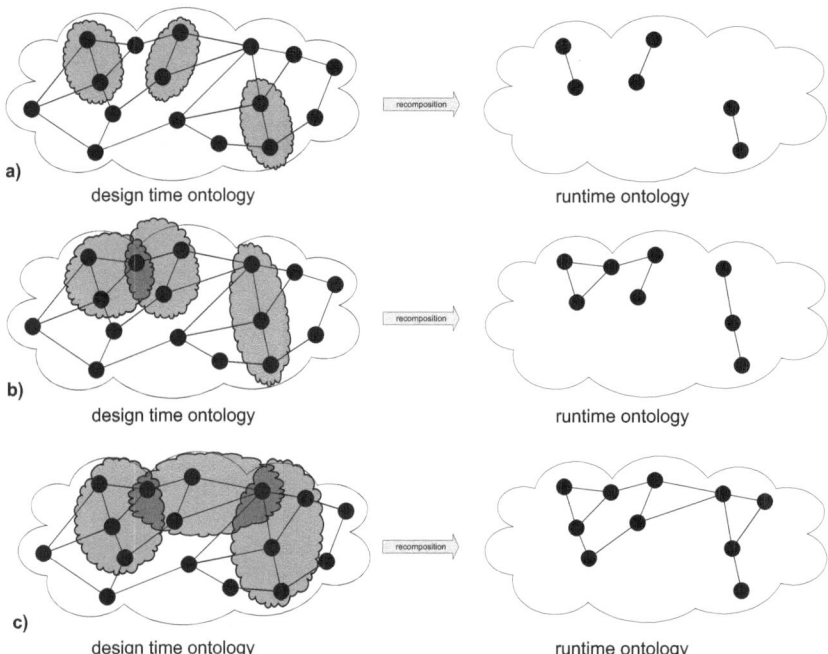

Fig. 5. The impact of service description richness. Although all descriptions share (and thus are connected by) model concepts, this connection is not employed in service matchmaking. Composing core descriptions only, might result in the runtime ontology taking the form of a set of isolated islands (a). Increasing the richness reduces this effect (b, c).

Increasing richness of the service descriptions, however, increases the size of the resulting ontology. Fig. 6 shows the impact of service description richness on the size of a RO. X-axis represents the number of service descriptions used in composing RO (e.g. 100% means that all service descriptions that were defined in the common ontology were used to compose RO). Y-axis shows the size of RO as a percentage of the number of concepts from the common ontology (e.g. 50% means that only half of the concepts from the common ontology were transferred to the RO).

It can be seen that using only core service descriptions ($r=0$) reduces the size of RO to 5-20% of the original ontology (depending on the number of description that were composed). With the increase of service descriptions' richness the RO becomes bigger (e.g. 22-46% for $r=2$, 68-78% for $r=8$). It is also interesting to note that the RO's sizes do not increase linearly with the increase of descriptions' richness. In fact, increasing description richness above 8 does not increase the size of the RO above 78% of the original ontology's size. The reason is the presence of concepts in the original ontology that are not used in service matchmaking. In the ontology used in this experiment they constituted 22% of all concepts. Ontology recomposition always 'filters them out' from the RO.

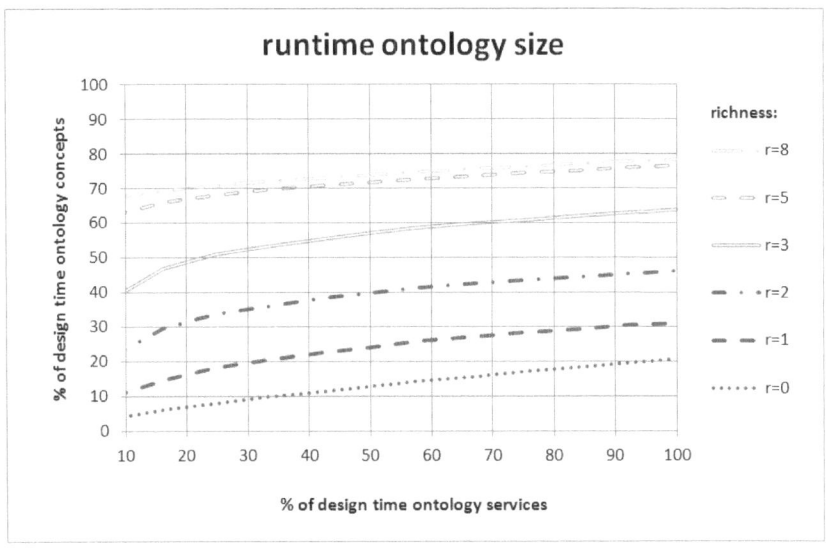

Fig. 6. The size of a RO composed from service descriptions of varying richness (r)

5.2   The Quality of Service Matchmaking

The reduction of the size of ontological representation of services must not be performed without considering its impact on its primal use: service discovery and compostion. Service discovery might be seen as a case of information retrieval and the impact of ontology recomposition on the service matchmaking may be measured with the use of the Cranfield model [45]. In order to assess relative performance of service matchmaking the results of matchmaking against runtime ontology are compared with the results obtained by performing the matchmaking against design time ontology consisting of the same service descriptions[2].

In this paper, the Recall measure from the Cranfield model is used. Recall is a proportion of documents that are retrieved to documents that are relevant to a query:

$$Recall = \frac{|A \cap B|}{|A|}, \qquad Recall \in\; <0; 1> \tag{1}$$

where B is the set of services obtained by resolving a query against a runtime ontology and A is a set of services obtained by resolving the query against a design time ontology that contains descriptions of the same services. Additionally, since RO is a subset of the common ontology, it is always true that B ⊂ A, the following formula may be used:

---

[2]   This is in accordance with a common sense - since a RO contains only a subset of a DO, a RO-based matchmaking for a given query will never deliver the same results. In order to be able to judge the quality of RO-based matchmaking, the services that are not present in the RO (and thus couldn't have been retrieved) need to be removed from the DO-based matchmaking results.

$$Recall = \frac{|B|}{|A|}, \quad Recall \in < 0; 1 > \tag{2}$$

Recall=1 means that the query against runtime ontology gives the same results when resolving against design time ontology.

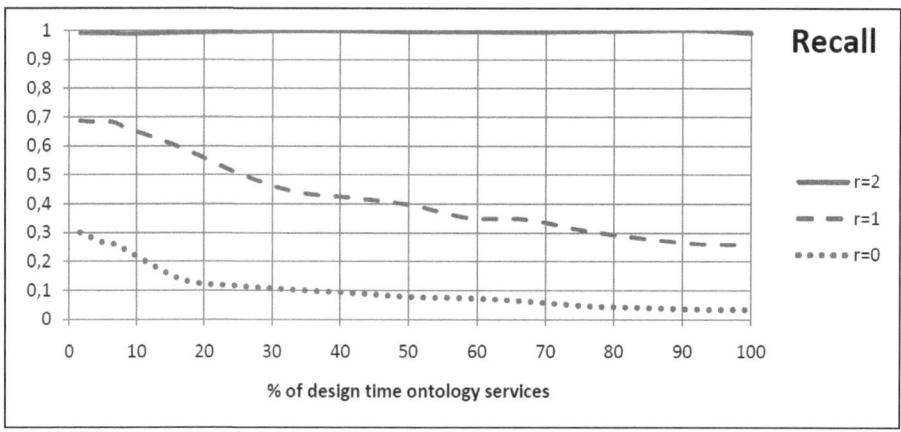

Fig. 7. Recall for SUMO-based runtime ontology

The recall measures shown in Fig. 7. were calculated by averaging individual recall values calculated by a set of 1934 different queries against a set of runtime 4503 runtime ontologies differing in number and richness level of their constituent service descriptions. X-axis represents the number of service descriptions used in composing RO (e.g. 100% means that all service descriptions that were defined in the common ontology were used to compose RO). Y-axis shows the averaged recall values. Unsurprisingly, ROs composed from core service descriptions offer very low matchmaking quality (Recall<0,3). The quality grows with the increase of description richness. With the ontology used in the experiment, using description richness of 2 allowed to obtain almost perfect matchmaking quality (Recall values ~1).

5.3   Assessing Sufficient Richness

In order to estimate the sufficient enrichment level of the constituent service descriptions the average distance between service descriptions needs to be calculated. In short, a distance between services is defined as a number of 'hops' in the common ontology between concepts representing their parameter types. If a distance between a service description and any other description in the RO is greater than their combined richness, the description becomes an isolated island (Fig. 5a). For instance, as Fig. 8 shows, richness 2 covers services whose distance between parameters is at most four hops: Srv_A with Param_7 and SrvB with Param_3 shown in Fig. 8a) lie within distance = 4. Also services whose parameters' distance is 3 require richness 2 to be composable (Fig. 8b): `Srv_A` with `Param_6` and `SrvB` with `Param_3` lie within distance=3. Descriptions of richness 1 does not cover the distance between `Param_4` and `Param_5` and thus render the composition impossible.

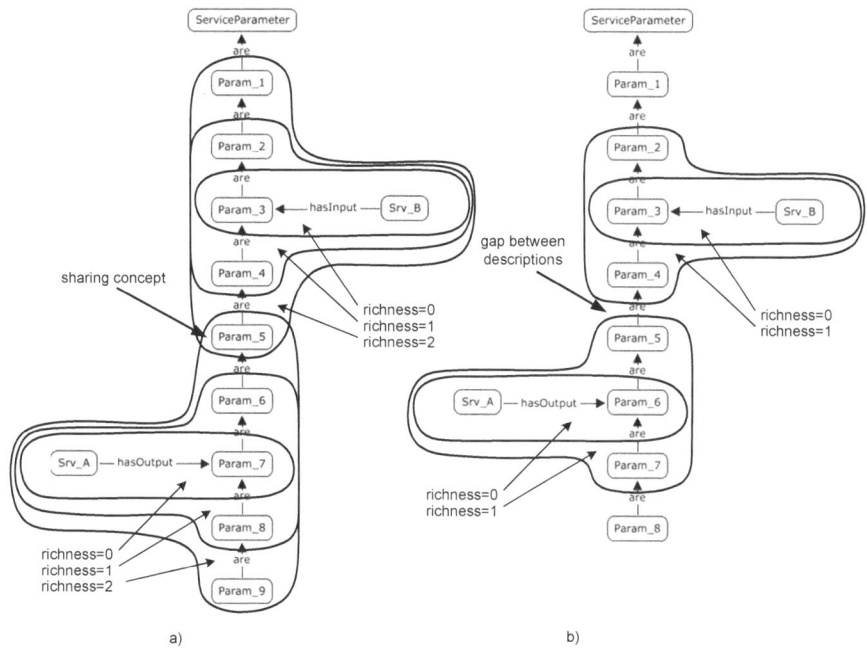

Fig. 8. Richness =2 equals the distance of four (a) or three (b) 'hops' in the parameter hierarchy tree

Tab. 1 shows the distribution of distances between composable (i.e. with finite distances) services in the common ontology used in the experiment. The table shows that that the distance between 94% of services is no greater than 4. This means that the richness of the service descriptions that was required for maintaining the quality of service matchmaking through the ontology recomposition is 2.

Table 1. Distribution of distances between services in ontology used in the experiment

Distance	0	1	2	3	4	5	6
% of services	13	30	28	15	8	4	2

By calculating the distribution of service parameter types of different services in a common ontology it is thus possible to assess the descriptions richness that is sufficient to provide good quality of service matchmaking. This way it is possible to give recommendations about the enrichment levels of the service descriptions that are to be used in a specific environment at design time, upon their creation and deployment in target devices.

6   Conclusion

Ontology recomposition technique allows to fine tune the ontology to a specific environment by reducing 'semantic richness' of descriptions of individual services present in an environment. The reduction level that is sufficient to maintain the quality of service matchmaking might be easily estimated from the original ontology.

References

1. Kashyap, V., Bussler, C., Moran, M.: The Semantic Web: Semantics for Data and Services on the Web. Springer-Verlag New York Inc. (2008)
2. Alonso, G., Casati, F., Kuno, H., Machiraju, V.: Web Services. Springer (2003)
3. Rybicki, T., Domaszewicz, J.: Sensor-Actuator Networks with TBox Snippets. In: Abdennadher, N., Petcu, D. (eds.) GPC 2009. LNCS, vol. 5529, pp. 317–327. Springer, Heidelberg (2009)
4. Boubez, T., Hondo, M., Kurt, C., Rodriguez, J., Rogers, D. (eds.) UDDI Programmer's API 1.0. UDDI Published Specification (2002), http://UDDI.org
5. Fernandez-Chamizo, C., et al.: Case-based retrieval of software components. Expert Systems with Applications 9(3), 397–421 (1995)
6. Fugini, M.G., Faustle, S.: Retrieval of reusable components in a development information system. In: Second International Workshop on Software Reusability. IEEE Press (1993)
7. Richard, G.G.: Service advertisement and discovery: enabling universal device cooperation. IEEE Internet Computing 4(5), 18–26 (2000)
8. Akkiraju, R., Goodwin, R., Doshi, P., Roeder, S.: A Method For Semantically Enhancing the Service Discovery Capabilities of UDDI. In: Proceedings of the Workshop on Information Integration on the Web, IJCAI 2003, Acapulco, Mexico, August 9-10 (2003)
9. Arnold, K., Wollrath, A., O'Sullivan, B., Scheifler, R., Waldo, J.: The Jini specification. Addison-Wesley, Reading (1999)
10. Aversano, L., Canfora, G., Ciampi, A.: An algorithm for web service discovery through their composition. In: Zhang, L. (ed.) IEEE International Conference on Web Services (ICWS 2004), pp. 332–341. IEEE Computer Society (2004)
11. Chakraborty, D., Perich, F., Avancha, S., Joshi, A.: DReggie: Semantic Service Discovery for M-Commerce Applications. In: Workshop on Reliable and Secure Applications in Mobile Environment, SRDS (October 2001)
12. McDermott, M.: Estimated-regression planning for interactions with Web services. In: Proceedings of the 6th International Conference on AI Planning and Scheduling, Toulouse, France. AAAI Press (2002)
13. McIlraith, S., Son, T.: Adapting Golog for composition of semantic Web services. In: Proc. 8th International Conference on Principles of Knowledge Representation and Reasoning (2002) (to appear)
14. Ni, Q., Sloman, M.: An Ontology-enabled Service Oriented Architecture for Pervasive Computing. In: The International Conference on Information Technology: Coding and Computing (ITCC 2005). IEEE Computer Society (2005)
15. Ni, Q.: Service composition in ontology enabled service oriented architecture for pervasive computing. In: Workshop on Ubiquitous Computing and e-Research (2005)

16. Benatallah, B., Hacid, M.S., Rey, C., Toumani, F.: Request rewriting-based web service discovery. In: Fensel, D., Sycara, K., Mylopoulos, J. (eds.) ISWC 2003. LNCS, vol. 2870, pp. 242–257. Springer, Heidelberg (2003)

17. Benatallah, B., Hacid, M.S., Leger, A., Rey, C., Toumani, F.: On automating web services discovery. VLDB Journal 14(1) (2005)

18. Mokhtar, S.B., Preuveneers, D., Georgantas, N., Issarny, V., Berbers, Y.: Easy: Efficient semantic service discovery in pervasive computing environments with QoS and context support. Journal of System and Software (2007)

19. Paolucci, M., Kawamura, T., Payne, T.R., Sycara, K.: Importing the Semantic Web in UDDI. In: Web Services, E-Business and Semantic Web Workshop (2002)

20. Paolucci, M., Kawamura, T., Payne, T.R., Sycara, K.: Semantic matching of Web services capabilities. In: Horrocks, I., Hendler, J. (eds.) ISWC 2002. LNCS, vol. 2342, pp. 333–347. Springer, Heidelberg (2002)

21. Sirin, E., Hendler, J., Parsia, B.: Semi-automatic composition of Web services using semantic descriptions. In: Proceedings of Web Services: Modeling, Architecture and Infrastructure Workshop in Conhunction with ICEIS 2003 (2002)

22. Sirin, E., Parsia, B., Hendler, J.: Filtering and Selecting Semantic Web Services with Interactive Composition Techniques. IEEE Intelligent Systems 19(4), 42–49 (2004)

23. Srinivasan, N., Paolucci, M., Sycara, K.: Adding OWL-S to UDDI, Implementation and Throughput. In: Proceedings of the First International Workshop on Semantic Web Services and Web Process Composition, USA (2004)

24. Klein, M., Bernstein, A.: Searching for Services on the Semantic Web using Process Ontologies. In: The First Semantic Web Working Symposium (SWWS-1), Stanford, CA, USA (2001)

25. Klein, M., Bernstein, A.: Searching for Services on the Semantic Web using Process Ontologies. In: The First Semantic Web Working Symposium (SWWS-1), Stanford, CA, USA (2001)

26. Seaborne, A., Prud"hommeaux, E.: SparQL Query Language for RDF (February 2005), http://www.w3.org/TR/rdf-sparql-query/ (accessed June 2009)

27. Alexaki, S., Christophides, V., Karvounarakis, G., Plexousakis, D., Tolle, K., Amann, B., Fundulaki, I., Scholl, M., Vercoustre, A.M.: Managing RDF metadata for community webs. In: WCM 2000, Salt Lake City, Utah, pp. 140–151 (October 2000)

28. Broekstra, J., Kampman, A.: SeRQL: An RDF Query and Transformation Language (2004)

29. Volz, R., Oberle, D., Studer, R.: Views for light-weight web ontologies. In: Proceedings of the ACM Symposium on Applied Computing, SAC (2003)

30. Magkanaraki, A., Tannen, V., Christophides, V., Plexousakis, D.: Viewing the Semantic Web through RVL Lenses. Journal of Web Semantics 1(4), 29 (2004)

31. Stuckenschmidt, H., Klein, M.: Structure-based partitioning of large concept hierarchies. In: McIlraith, S.A., Plexousakis, D., van Harmelen, F. (eds.) ISWC 2004. LNCS, vol. 3298, pp. 289–303. Springer, Heidelberg (2004)

32. Grau, B.C., Parsia, B., Sirin, E., Kalyanpur, A.: Modularizing OWL Ontologies. In: K-CAP 2005 Workshop on Ontology Management (October 2005)

33. Batagelj, V.: Analysis of large networks - islands. Presented at Dagstuhl Seminar 03361: Algorithmic Aspects of Large and Complex Networks (2003)

34. Seidenberg, J., Rector, A.: Web Ontology Segmentation: Analysis, Classification and Use. In: Proc. of the World Wide Web Conference (WWW) (2006)

35. d''Aquin, M., Sabou, M., Motta, E.: Modularization: a Key for the Dynamic Selection of Relevant Knowledge Components. In: Proc. of the ISWC 2006 Workshop on Modular Ontologies (2006)
36. Pinto, S., Gomez-Perez, A., Martins, J.: Some Issues on Ontology Integration. In: Proceedings of the IJCAI 1999 Workshop on Ontologies and Problem-Solving Methods (KRR5), Stockholm, Sweden (August 1999)
37. Choi, N., Il-Yeol, S., Han, H.: A Survey on Ontology Mapping. ACM SIGMOD Record 35(3), 34–41 (2006)
38. Klein, M.: Combining and relating ontologies: an analysis of problems and solutions. In: Proceedings of the IJCAI 2001 Workshop on Ontologies and Information Sharing, Seattle, WA (2001)
39. Shvaiko, P., Euzenat, J.: A Survey of Schema-based Mapping Approaches. Technical Report, DIT-04-087,University of Trento (2004)
40. Rahm, E., Bernstein, P.A.: A Survey of Approaches to Automatic Schema Matching. The VLDB Journal 10 (2001)
41. Do, H.-H., Melnik, S., Rahm, E.: Comparison *of Schema Mapping Evaluations.* In: Proceedings of the GI-Workshop "Web and Databases", Erfurt (2002)
42. Xiao, L.S., Ellen, R.: Automated Schema Mapping Techniques: An Exploratory Study. Res. Lett. Inf. Sci. 4, 113–136 (2003)
43. Doan, A., et al.: Ontology Mapping A Machine Learning Approach. In: Staab, S., Studer, R. (eds.) Handbook on Ontologies in Information Systems. Springer (2003)
44. Rybicki, T., Domaszewicz, J.: Dynamic service discovery and composition in networks of resource-constrained nodes. In: Krajowe Sympozjum Telekomunikacji i Teleinformatyki (KSTiT 2009), Warsaw, Poland (September 2009)
45. Voorhees, E.M.: The philosophy of information retrieval evaluation. In: Peters, C., Braschler, M., Gonzalo, J., Kluck, M. (eds.) CLEF 2001. LNCS, vol. 2406, pp. 355–370. Springer, Heidelberg (2002)

Structured Construction of Knowledge Repositories with MetaConcept

Germán Bravo

Universidad de Los Andes, Departamento de Ingeniería de Sistemas y Computación,
Kr. 1 Este # 19ª-40, Bogotá, Colombia
gbravo@uniandes.edu.co

Abstract. One of the main issues during the development of a knowledge repository for the Technical Area of Architecture, called KOC, was the organization and structuration of this knowledge domain: the ArCo ontology. Although powerful tools to write ontologies exist, they do not guide their development. MetaConcept answers this need by defining the concepts that describe the structure of the ontology and its construction process: first a meta-ontology for the domain, then the ontology itself and then the knowledge repository. Two examples are presented, KOC 2.0, a data repository for real-world experiences in building construction and its knowledge repository, and ELLES, a repository of lessons learned in the domain of software construction projects.

Keywords: Ontology, meta-ontology, knowledge repositories, knowledge objects, Architectural Ontologies.

1   Introduction

The project KOC [1] [2] [14], conducted by the COMIT and ACE Groups of University of Los Andes, aims to generate a repository of real-world examples of the application of building construction techniques. The goal is to allow instructors and students to gain access to information concerning the dynamics of a construction site without actually having to organize visits, thus avoiding safety hazards and problems concerning the availability, scheduling and coordination of the parties involved.

This paper presents MetaConcept, one of the conceptual results of the project, which has led to a mechanism to build purpose specific ontologies aiming the definition of data and knowledge repositories with educational purposes. Initially designed and intended for the field of building architecture, MetaConcept extracts the ontology fundamental structure and constitutes a meta-ontology applicable to other domains of knowledge. Although the concepts and examples presented are mainly drawn from the field of architecture, an example of application for a lessons-learned system is also shown.

For the subject of ontologies and ontology construction, a substantial amount of literature exists [12] [6] [10] showing descriptions of several tools pursuing the same objectives as MetaConcept. Without prejudice to other applications, some of these are Methontology [4] [5], Ontology Seeker [7] and Protegé [11]. The main concern

M. Graña et al. (Eds.): KES 2012, LNAI 7828, pp. 133–142, 2013.

regarding these applications is that they do not provide guidelines to orient the conceptual development of ontologies, allowing for too much freedom that leads to a possible lack of structure, making them difficult to understand and fully exploit.

Regarding the Architecture domain, stands the MACE portal [8]: it supports ontology-based searches for European architectural projects, but focuses on finished projects rather than on the architectural techniques used to build them.

The present article starts by laying the conceptual groundwork for MetaConcept, and by defining the stages in the construction of a knowledge repository. Furthermore, two examples show how MetaConcept has been used, including the software architecture of KOC 2.0, which allows the generation of three different types of products. The article then provides some suggestions to effectively exploit the semantic power of repositories built with MetaConcept, and finally some comments and conclusions are drawn.

2   Construction of Knowledge Repositories

The main objective of MetaConcept is to guide and support the construction of knowledge repositories aimed to facilitate the teaching of a knowledge domain by: (1) giving a fully structured description of concepts in the domain; (2) the creation and browsing of Knowledge Objects (KO) showing real life examples of the use and misuse of these domain concepts; and (3) providing semantic searches of interest for instructors and students.

The conceptual process to create a knowledge repository comprises four stages. Section 3 shows a detailed example of expected results and section 4 shows some alternatives to exploit these knowledge repositories.

2.1   Stage 0 – Global Structure of MetaConcept

Based on [10], MetaConcept defines the conceptual structure of a knowledge domain in five elements, as shown in Figure 1.

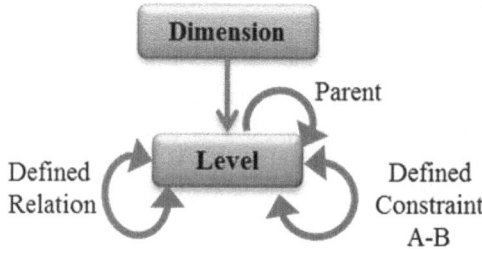

Fig. 1. Main structure of MetaConcept

- The Dimensions define the conceptual framework of the domain, those features and characteristics that which cannot be disregarded when describing an object in the knowledge domain. Each dimension, regardless of its complexity, should be described and studied independently.

- The Levels of Detail complete the description of a dimension. Each dimension is described by a hierarchical list of levels. The relationship between two consecutive levels may be of inheritance (is-a) or composition (part-of).
- The Defined-Relationships relate two dimensions by establishing the appropriate level of detail in which they interact. A defined relationship is a tuple <n1, d1, n2, d2> where n1 is a level of dimension d1 and n2 is a level of dimension d2, indicating that any concept of n1 may relate to any concept of n2.
- The Defined-Constraints define relationships between two dimensions restricted by rules of coherence and integrity. A defined constraint is a tuple <n1, d1, n2, d2>, where n1 is a level of dimension d1, and n2 is a level of dimension d2, indicating the level of detail of the two dimensions subject to the constraint. There are two types of constraints: by inclusion (A) representing the truths of the domain, i.e., what "should be"; and by exclusion (B) representing only the valid combinations of concepts, i.e. the "only these": the inconsistencies and prohibitions in the domain are its complement, i.e., what "cannot be".

Any knowledge domain can be structured using these five elements.

2.2   Stage 1: Construction of a Meta-Ontology

Stage 1 searches the expression of the conceptual structure of the knowledge domain at hand in terms of MetaConcept elements: The dimensions, levels, defined-relationships and defined-constraints must be identified and organized. The result, although not fully detailed, presents a comprehensive overview of the domain: The goal is that any concept in the knowledge domain has a place (and only one place) in the conceptual framework defined by this meta-ontology.

2.3   Stage 2: Construction of the Ontology

Stage 2 corresponds to the actual construction of the ontology by instantiating the specific domain concepts and the specific restrictions of the meta-ontology.

Each dimension is described as a tree of concepts, following the hierarchical relationships established in the meta-ontology. Once the dimensions have been defined, it becomes necessary to define the constraints between them, according to those defined in the meta-ontology: given two concepts of two different dimensions, a restriction is created if and only if the meta-ontology includes a defined-constraint between the respective levels of corresponding dimensions.

The result is shown Figure 2. Dotted arrows represent the fact that relationships instances defined in the ontology respects the relationships defined in the meta-ontology.

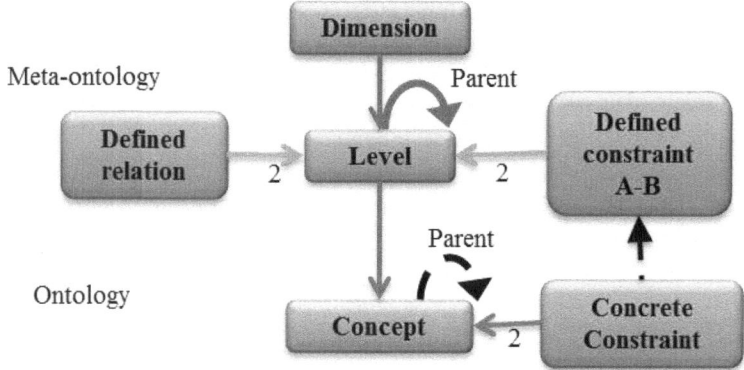

Fig. 2. Ontology Structure

2.4   Stage 3: Population of the Repository

Given the ontology, Stage 3 corresponds to the creation and population of a repository of knowledge objects (KO) containing, for every concept in the ontology, real life application examples drawn from the domain at hand. Every KO is described (annotated) according to the semantics defined by the ontology and scored according to its validity (is it a good or a poor practice?). Figure 3 shows the expected result, where the dotted arrows indicate that every annotation is supported by the existence of a Defined Relationship in the meta-ontology and must not violate any of the specific constraints defined in the ontology.

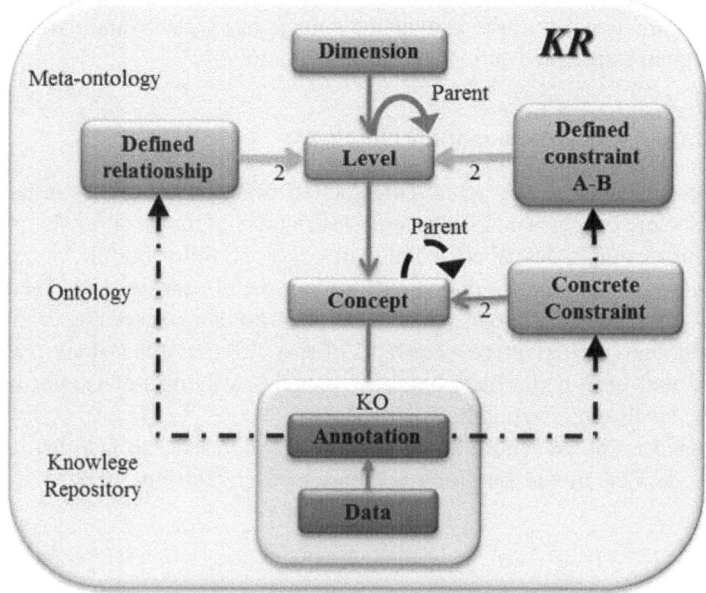

Fig. 3. The Global Picture: A knowledge repository built with MetaConcept

It is important to notice that while the meta-ontology must be complete and coherent, the ontology thus defined is consistent and ready to use, but its completeness is not compulsory: it is not necessary to count with all the concepts in all dimensions. In any case, the ontology can be used either for pedagogical purposes, or to create new knowledge objects. The ontology and knowledge base should evolve simultaneously, in a coherent manner.

3 Application Examples

3.1 KOC 2.0 – Knowledge Objects of Construction

The first example shows the ArCo ontology [14] that was used to build KOC [1], a knowledge repository about building construction.

Issued from Stage 1, Figure 4 presents the meta-ontology ArConcept, the conceptual structure of the knowledge domain of building construction, constructed by instantiating the macro concepts of the architecture domain in terms of the elements provided by MetaConcept:

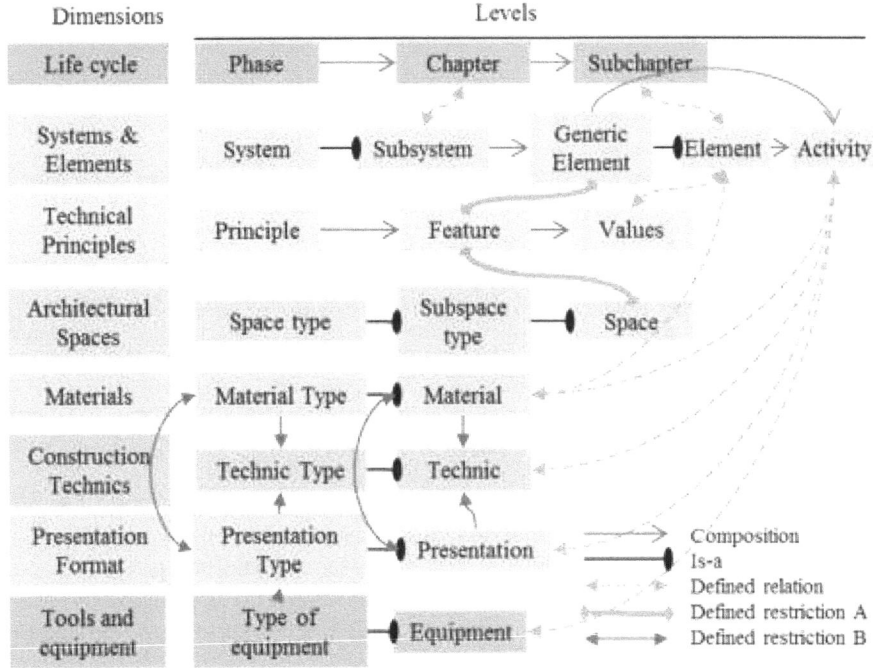

Fig. 4. ArConcept

- Dimensions: The domain of building construction has eight dimensions. A building has a life cycle; defines a set of spaces; is composed by a set of systems; must be built by following the architectural technical principles; there exist techniques and equipment appropriate for materials used, which can arrive in different formats to the construction site.

- Levels: Each dimension has several levels of detail. In particular, the Systems and Elements dimension establishes that the building has several Systems, subdivided into Subsystems; the subsystems comprise several Generic Elements, which may be constructed in diversity of forms and materials, following a sequence of Activities.
- Defined-relationships occur at the lowest levels of dimensions. For example, the relationship between the Systems and Elements and Material dimensions occurs at the Element level. It is important to note that Activity constitutes the greatest detail level connecting all the dimensions in the domain.
- Defined-constraints(A): There are two of these restrictions, and they establish precisely that the Elements and Spaces of the building must follow the basic Technical Principles of the Architecture
- Defined-constraints (B): There are several of these constraints, representing the fact that the Techniques and Equipment used in the Activities depend on the Materials used and the Format in which they arrive to the construction site.

Stage 2 leads to ArCo, an ontology for the technical domain of the Architecture, specifically about building construction, created according to the structure defined in ArConcept. Figure 5 exemplifies the Systems and Elements dimension, with two concepts in each level.

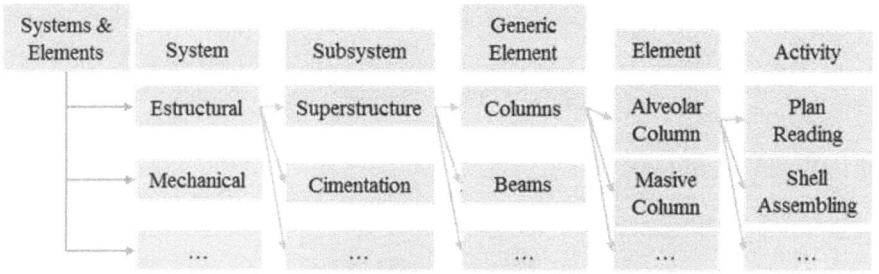

Fig. 5. ArCo excerpt

As stated earlier, MetaConcept arises as part of the evolution of the KOC project, whose initial version can be found in [13] and allowed the construction of a new version, KOC 2.0, which exploits the expressive power of MetaConcept.

The general software architecture of the application, Figure 6, allows the construction of three types of knowledge applications, each one representing a different product, with its own advantages, applications and users.

1. ArcConcept and ArCo constitute a complete application based on MetaConcept aiming the creation, editing, and browsing of the ontology and the meta-ontology of the knowledge domain of building construction.
2. DataConcept and Metadata also constitute a complete application based on MetaConcept aiming the creation, editing, and browsing of the ontology that describes

the metadata of the architectural data. It includes "classic" document metadata (format, date, author, etc.), and also metadata for the architectural domain, such as the names of the architects involved, and the name and location of the project, among others.

3. Data refers to the actual architectural data repository (maps, images, videos, etc.), described through its metadata ontology. The user interface allows creating, editing, and navigating the data.

4. KOC is the knowledge repository of building construction. It is built based on the previous products and allows the creation, editing, and navigation of knowledge objects (KO). This level has an auxiliary application, BarKO [3], which provides a semantic query service and ordered navigation of KOs. The queries are made based on ArCo and Metadata ontologies; responses are displayed, by default, respecting a semantic ordering of dimensions in the domain: life cycle, systems and elements, spaces, technical principles, materials, types of presentation, techniques and equipment.

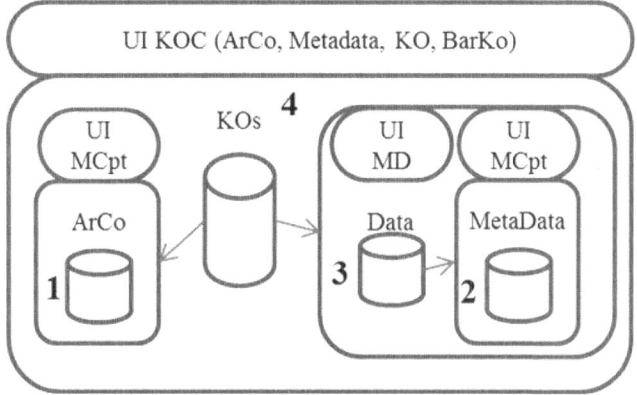

Fig. 6. Architecture of KOC 2.0.

Again, it is important to note that once you have ArConcept and DataConcept, the system can evolve in parallel at all levels: ArCo and Metadata can be created or completed, and the data and knowledge repositories can be created and browsed.

3.2   Elles – Enterprise Lessons Learned System

Another example of the use of MetaConcept is Elles [9], a lessons-learned repository for the specific context of technology projects. The knowledge domain comprises the stages and deliverables of a software development project, to which the lessons-learned are referred. Following the same conceptual process and development used in KOC, the following ontologies are built: LeConcept, equivalent to ArConcept; and Leleo, equivalent to ArCo.

The main difference between these two projects, besides the knowledge domain, lies in the characteristics of the knowledge objects (Figure 7). A lesson-learned can refer to various aspects of the project at different times: using Leleo, a lesson learned captures a situation as soon as a problem is detected, as well as its possible causes and the actions taken to solve the problem. Each element of the lesson learned may also refer to a different stage or deliverable of the software project.

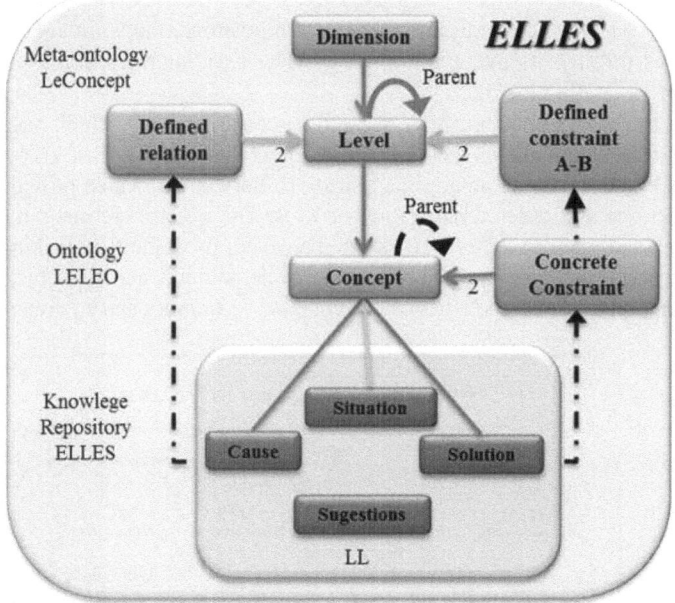

Fig. 7. Lessons Learned Repository built with MetaConcept

4   Exploitation of the Knowledge Repository

Once the domain has been structured, and the knowledge repository generated, it is possible to define pedagogical activities and also build some other interesting applications profiting the structure and content of the repository. These are explained using ArConcept, ArCo and KOC, but are equally valid for other repositories:

- Browsing the meta-ontology and the ontology supports the transmission of the structure and concepts of the knowledge domain.
- Browsing of KOs allows the critical and comparative analysis of real-world application examples of the concepts. The browsing of the KOs profits from the relations and restrictions defined in ArConcept to give complete and consistent information about the KO; Barko solves the problem of the multiplicity of responses by showing the KOs in a semantically ordered fashion.
- The creation of KOs by users and students forces them to understand and adhere to the structure, relationships and constraints in the knowledge domain.
- Advanced query processing takes advantage of relations and constraints in the ontology. This allows knowledge to be inferred from actual data contained in the

repository as, for instance, good and bad practices, weaknesses and strengths of the company.

- The connection of the repository with other tools for ontology management and inference engines, profiting from the exportation of the ontology to OWL.
- Connection of the repository with tools for data mining, which can discover hidden patterns in the data and in the KO repository.
- And many more...

5   Concluding Remarks

The main point is that MetaConcept effectively achieved its intended goal by providing a framework to the conceptual development of ontologies, without sacrificing expressive power. The organization of a knowledge domain into dimensions and levels allows the expression of the domain at different levels of detail facilitating its construction and understanding. In summary:

- Different levels of abstraction facilitate the construction of ontology-based data repositories and knowledge repositories comprising several semantic levels in an organized fashion.
- It is possible to compose several knowledge domains and build multi-ontology applications, as shown by KOC which integrates ArCo, Metadata and the KOs.
- It is possible to generate a uniform interface for the creation of ontologies (i.e., its structure and the concepts of the domain), the creation of KOs, ontology browsing, KO browsing and queries (BarKo).
- The dynamism of the ontology and data and knowledge repositories is guaranteed. Ontology, metadata and knowledge objects share the structure of the knowledge domain and can evolve independently without complications.
- MetaConcept repositories allow data to present their meaning, with a user-friendly interface: all of the relevant structural information is always available to the user.

All of the above constitutes the many and varied didactic advantages of the present proposal, for the teacher, the KO creator, for students, and for people interested in the knowledge domain at hand.

Acknowledgments. To the architects Rafael Villazón and Augusto Trujillo, with whom the different versions of ArConcept and ArCo have been developed; to the engineers Luis Javier Bautista, Nelson Cruz, Héctor Sánchez, Carlos Omaña, Mauricio Caviedes and David Cifuentes, by the software development.

References

1. Bravo, G., Villazón, R.: KOC - Manejo Ontológico de Objetos de Conocimiento en la Construcción de Edificios. In: Avances en Sistemas e Informática, Medellín, vol. 4 (2007)

2. Cifuentes, D.: KOC: Un Repositorio de Objetos de Conocimiento Para la Enseñanza de la Construcción - Tesis de Magíster en Ingeniería de Sistemas. Universidad de los Andes, Bogotá (2007)

3. Cruz, N., Bravo, G.: BArKo: Buscador Navegador de Objetos de conocimiento Descritos por ArCo en el proyecto KOC. In: Latin American Conference on Networked and Electronic Media, Cali (2010)

4. Gómez-Pérez, A., Corcho, O.: Ontology Languages for the Semantic Web. IEEE IntelligentSystems 1(17) (2002)

5. Gómez-Pérez, A., Fernández-López, M., Corcho, O.: Ontological Engineering: with examples from the areas of knowledge management, e-commerce and the Semantic Web. Springer, New York (2003)

6. Horridge, M.: Protégé OWL Tutorial. The University of Manchester (2011), http://owl.cs.manchester.ac.uk/tutorials/protegeowltutorial

7. Lim, E., Liu, J., Lee, R.: Knowledge Seeker – Ontology Modelling for Information Search and Management. Springer, Heidelberg (2011)

8. Metadata for Architectural Contents in Europe, http://portal.mace-project.eu/Home

9. Numpake, A.: ELLES – Enterprise Lessons Learned System. Tesis de Magíster en Ingeniería de Sistemas. Universidad de los Andes, Bogotá (2012)

10. Schwarz, U., Smith, B.: Ontological Relations. In: Reicher, M.E., Seibt, J., Smith, B., Wachter, D.V. (eds.) Applied Ontology: an Introduction. OntosVerlag, Heusenstamm (2008)

11. Stanford Center for Biomedical Informatics Research: Protegé, http://protege.stanford.edu/

12. Subhashini, R., Akilandeswari, J.: A survey of Ontology Construction Methodologies. In ternational Journal of Enterprise Computing and Business Systems 1(1) (2011)

13. KOC Web page, http://157.253.201.47:8080/KOC/administracion/inicio.html

14. Villazón, R., Bravo, G., Cifuentes, D.: ArCo: An Ontology for Architectural Concepts in Construction. In: Knowledge Generation, Communication and Management: KGCM-15, Orlando (2008)

Association between Teleology and Knowledge Mapping

Pavel Krbálek and Miloš Vacek

University of Hradec Králové, Faculty of Informatics and Management,
Rokitanského 62, 500 03 Hradec Králové, Czech Republic
{pavel.krbalek,milos.vacek}@uhk.cz
http://www.uhk.cz/fim

Abstract. Knowledge mapping can be performed in many different ways on the web, today. One of them, and increasingly popular, are social networks where people can share various content among friends, in user groups or with public. However, the majority of these networks are for socializing and entertainment. To emphasize the work and results we use teleology, a branch of philosophy, that claims that each object has a final purpose. In this paper, we introduce a framework for knowledge mapping where teleological perspective of knowledge is mapped and pre-defined ontology provides its context. Thus, we are able to design a social network for strategic knowledge mapping, useful to track objectives and areas of interest in academic as well as in business environment.

Keywords: Teleology, Knowledge mapping, Strategic Knowledge Map.

1 Introduction

Teleology in philosophy applies to any system attempting to explain a series of events in terms of targets, goals or purposes. In this article, we show how teleology is used as a modern scientific discipline that provides a channel to achieve different goals. In more detail, it focuses on using teleology in knowledge mapping where it serves as a framework for capturing strategic knowledge. An important part of this in a knowledge map is setting individual objectives. However, without some meaningful context, the different knowledge types are hardly distinguished. Therefore, we introduce ontology for mapping these knowledge types separately, in academic environment, through assigning folksonomy tags to knowledge nodes. Further, we suppose that suggested ontology can be helpful in profit or non-profit organization where there are needs to improve cooperation between individuals, existing teams or departments.

Knowledge management as a discipline itself includes several subbranches, each of which deserves its attention. Knowledge management processes are based on philosophies of Aristotle or Plato. The concept is based on best practices from philosophy, information system management, organization science, management science, cognitive science and other domains. Understanding and using knowledge management methods and tools means understanding simple principles,

M. Graña et al. (Eds.): KES 2012, LNAI 7828, pp. 143–152, 2013.
© Springer-Verlag Berlin Heidelberg 2013

to which often leads a difficult journey. Viewing this from the perspective of the knowledge hierarchy, with data at the basement and rising over information and knowledge, to the wisdom on the top, then managing information is the most widespread in the business environment. Knowledge management is understood in various ways but often lacks the weight it deserves. In what way should managers use knowledge management systems as appropriate assistance? Where possible, aiming at knowledge, its localization, exchange and storage, is the correct answer. The processes of knowledge management cannot be simplified only to storage (e.g. wiki solutions, expert systems), but it starts from the knowledge owner and the accessibility. Concentration on the availability and traceability of knowledge appears to be the first important step of knowledge management and knowledge mapping (KM) can serve as a tool to point out useful resources.

The structure of the rest of the paper is as follows. We first define basic terms knowledge types and introduce overview of research on knowledge mapping, knowledge maps categorization and knowledge mapping disadvantages. The section 2.3 introduces collaborative tagging as a useful technique that eliminates some disadvantages of others and can be enriched by semantic ontologies. Summary of teleological approach and its relation to presented knowledge mapping techniques and knowledge types is described in section 3. Finally, proposed knowledge mapping approach, based on teleology and strategic knowledge relation to all knowledge assets, is applied on academic environment.

2   Knowledge Management Theory

2.1   Knowledge Types

In general, we must be clear in what we mean by knowledge in knowledge mapping. This chapter brings a short overview of basic terms. The common view of knowledge, knowledge type and knowledge mapping is still missing as definitions of the individual authors also differ. Michael Polanyi, cited by [1], was the first who suggested two knowledge types, explicit and tacit. Tacit knowledge is essential to manage any human activity, particularly business processes. It contains organized and useful data but logic of its organization is too complex and ambiguous. Tacit knowledge is something that we know but are unable to express. Stories, rules or math algorithms are examples of explicit knowledge that can be stored in documents, manuals or databases. Cognitive psychologists defined descriptive (know-what) and procedural (know-how) knowledge [2]. Descriptive knowledge defines a state of the world, description of objects, situations, facts or methods and procedures [3]. Descriptive knowledge is also called declarative and is more comparable with explicit than tacit dimension. Procedural knowledge specifies doing something, actions or manipulations, for example, steps to fulfill a task. Strategic knowledge (know-why, know-when) is considered as the third type. It is rarely measured and is invoked only when other knowledge types are used [4]. A person in a decision process considers strategic knowledge to find the best solution. Each of their decisions is related to an objective and the sequence of decisions leading to the objective can be called a strategy. Strategic knowledge can be considered as a subset of declarative knowledge.

[5] states that knowledge becomes static when the tacit type is made explicit. Nowadays, investment into knowledge expression is not too important. Instead, resources of right knowledge accessible at the right time are of much higher importance. Also, knowledge mapping techniques can significantly improve knowledge traceability.

2.2   Knowledge Mapping

Knowledge is a fundamental part of a company but it cannot be captured sufficiently with a common formal structure. Common organization structure (organization chart) provides only names and positions of employees. In contrast, knowledge maps typically point to people, their experience as well as to documents and databases. [6] noted that developing a knowledge map involves locating important knowledge in an organization and then publishing some kind of a list or a chart that shows where to find it.

The key to understanding why someone should use this map is to understand how people learn. We usually learn by associating objects with other objects. Thinking about one topic, in individual's mind, triggers seeking another topic in the web of their thoughts. This philosophy concept is known as associationism and its history goes back to Plato, Aristotle or Kant [7]. One thought triggering another creates a network of interrelated nodes and results in knowledge creation. Associationism approach is also used in computer systems and is equivalent to hyperlinking.

There are various ways how knowledge maps can be classified. In the context of knowledge management, knowledge map brings an overview of knowledge related content elements: experts, project teams, communities, papers, patents, events, applications and so on. [8] presents that knowledge maps can be divided into three knowledge models: (1) charts such as hierarchical diagrams, (2) network diagrams that show nodes connected by arrows, (3) tables and grids, e.g. forms, matrices, frames or timelines. [9] focuses on cognitive maps and defines different categorization: text and language analysis maps, network maps, conclusive maps, classification maps and schematic maps.

Knowledge mapping techniques were categorized also by Caldwell into three basic groups [10]:

- **Procedural knowledge maps** - show knowledge and sources of knowledge mapped to business processes. The map consists of all relevant business processes where knowledge can be input or output.
- **Conceptual knowledge maps** - help to structure concepts and terms, to classify the content and taxonomy and to discover concepts and ontologies.
- **Social knowledge maps** - show relationships between individuals, skills and positions and help to find required expertise in people within an organization. Social network analysis and other techniques support finding the right knowledge owners.

[11] summarized different classification principles in one matrix. Different knowledge map classifications are based on knowledge management processes, content on knowledge maps, application levels, graphic forms and creation method types.

Knowledge Mapping Disadvantages. Basic disadvantages of knowledge maps identified by [3] can be summarized as: Misinterpreting, Information overload, Outdating, Unavailability and Costs of creating. When knowledge mapping is realized as a statistic survey, it is time consuming. Usually a specific relevant tool is required as well as knowledge of the terminology and notation (e.g. Topic Maps [12]). Mapping knowledge is also very tedious and people usually feel a threat that either it reveals that they know nothing (but keep pretending) or they are monopolists.

2.3   Collaborative Tagging

This chapter describes collaborative tagging as a type of conceptual knowledge maps. Next, we mention ontologies as a useful set of restrictions that provide meaning for tags assigned by users. There are different sources of information to be found in existing Web 2.0 applications. Users can store references to these sources and links to online resources can be shared and extended with other user information [13]. *Folksonomy* or a collaborative tagging is a way how to make order in such a complicated structure of resources. Folksonomy is meant to classify content in a computer network [14] using tags. The term itself is an analogy to the word *taxonomy*, where the Greek word *taxis* means *order* or *arrangement*, the word *nomos* means *law*. English *folks* is a derived term that can be explained as *people* [15]. Folksonomy offers a solution for the representation of resources by individual users in the virtual world and does not specify any rules. The principle lies in assigning any label called a *tag* to web pages, photos, documents and other online resources.

A relevant question is if folksonomy can be considered a knowledge map. The documents that are necessary to input and output from business processes contain explicit knowledge that can be added to a company folksonomy. [16] stated that social classification generated by employees reflects the real situation of knowledge understanding. First of all, tagging can provide an insight into individual's expertise and facilitate learning from others. The social dimension of the document tagging process is creation of social networks. Social network structure consists of interconnected individuals, who have similar interests, activities or personal relationships. If two documents are marked with the same tags then links between documents can be found as well as links between users who assigned them [17].

2.4   Ontology

A collaborative online network was successfully applied in many progressive firms such as Oracle or IBM. Two examples of social networking tools are Oracle Web-Center [18] or IBM Connections [19], internally developed by significant leading

IT companies. Adding tags and using tagcloud to locate people and documents are essential functions of such tools. On the other hand, defined ontology for tags, to add semantics, is not common. There is a large amount of studies concerning creation of more effective folksonomies by semantic meanings of words defined in the ontology or general principles of Nature language processing [20]. Ontology is defined as an explicit specification of a conceptualization [21]. Ontologies support knowledge sharing and represent vocabulary and relations in a specific domain. Benefits of using the ontology are that one knowledge management system can communicate with others or that systems retrieve automatically more precise collection of knowledge. [22] presents a way how to enable searching in various publications from various institutions. Folksonomies content is dependent on users; there are no restrictions to define tags.

The main weakness of folksonomies is uncontrolled vocabulary. [23] summarized disadvantages such as acronyms, synonyms, homonyms. For example, acronym "ANT" can be a shortcut for "Actor Network Theory" from the sociology domain or "Apache Ant" software building tool. A possible way to reduce these shortcomings is to establish some boundaries for what tags a user can insert into the system. Different theories how to bring additional semantics into folksonomies were published, e.g. in [24], [25]. An example of teleology approach as an ontology class in academic environment is given in section 3.2.

3   Knowledge Mapping Approach Based on Teleology

There are three key contributions of this paper. The first is to show how strategic knowledge is related to teleological perspective, in knowledge mapping. The second is the introduction of a new knowledge mapping approach based on teleology which allows individual's objective to be specified and externalized. The example describes academic environment and related model with ontology classes. Third contribution is a practical approach, based on collaborative tagging, which enables to specify a finite set of concepts and to create company's strategic knowledge map.

3.1   Teleology Perspective

This section describes philosophical background of teleology. Understanding philosophical base of knowledge enables us to move to relevant knowledge management processes. In Aristotle's Nicomachean Ethics there are presented five virtues, that can be mapped to knowledge levels. Aristotle's types are used to cover all possible acts of knowing [26]:

- Scientific knowledge *episteme*,
- Skill based knowledge *techne*,
- Practical wisdom based on experience *phronesis*,
- Intuition *nous*,
- Theoretical knowledge of universal truths *sophia*

In the following, we will concentrate on techne and phronesis. Whereas techne involves practical rationality, phronesis is characterized as reasoning, pragmatic, context-dependent and oriented toward action. Phronesis comprehend the ability to determine how to reach a certain target.

The effectiveness and purposefulness characterize human behavior and distinguishe it from purposeless reactions. Aristotle argued that all objects have a definite purpose [27]. Teleology in philosophy applies to any system attempting to explain a series of events in terms of targets, goals or purposes. [28] provides natural examples of teleology such as heart is for pumping blood or eyes are for seeing. It is opposed to mechanism, the theory that all events may be explained by mechanical principles of causation. [29] stated teleological tendency that knowledge is always there for a purpose towards some objectives. Every kind of live being naturally follows its target. This purpose is to live or have children but natural to humans is happiness. Against this, naturalism stands in opposition that subject does not foresee a predetermined goal. [30] uses teleological representation in robots navigation. Purposes are the abstraction of states of reality. Thus, the goal can be used as a relation between a robot and the physical environment.

The concept of purpose can not be easily formalized but it is clearly connected with strategic knowledge. Strategic knowledge concerns the specific purpose for human effort to reach an objective. Definition of relation between strategic knowledge and other knowledge types can be found in section 2. Strategic knowledge [31] is the key asset of organization. The strategic knowledge of a person who wants to achieve a goal in certain area is associated with tacit knowledge. Tacit knowledge is difficult to express, the least that can be said is where it is and how it is used. Thus, strategic knowledge map can serve as a structure where tacit knowledge is embedded. According to [32], an organization must be seen as a system in which the workers share meanings and beliefs (strategic knowledge). Sharing of individual's objectives (knowing-why) and associations between them can be made more effective through information technology, with collaboration tagging in particular. Knowledge mapping should be performed continuously along with everyday operations; it is not a single-use process. One of the main purposes of business administration should be to maintain meanings and beliefs at their best shape, to make organizational culture look complex, consisting of knowing *why*, knowing *who* and knowing *what* dimensions.

- **Knowing who** - an individual is identified with an objective, a single person or a group can benefit from objective sharing,
- **Knowing what** - electronic documents connected with some objective include important explicit knowledge (facts),
- **Knowing why** - an objective has its importance within a group of people, people discuss about different objectives of their work together. Explicit objectives publication can improve collaboration in a community.

3.2   Academic Knowledge Network

The main goal of teleology approach, presented in this paper, is mapping individual or community goals and creating links between them. Using collaborative tagging enables mapping user's objectives of their common work as a dedicated tag in a folksonomy. An objective represents individual or group topic of interest, e.g. each academic paper has a relevant research objective. In Fig. 1, we are presenting basic ontology model for a collaborative filtering tool that enables sharing research and academic content. University employees, PhD students and Master's degree students should have access to a common social network where each of them could create, update and share their research objectives, create or join in public and private groups according to areas of interest, share ongoing work and final results of research activities, information about planned events and more.

Domain ontologies are available for a variety of different domains, few of them were created specifically for the academic environment. AKT Reference Ontology [33] with 160 classes is useful for representing an academic computer science community. AIISO ontology [34] describes organization structure of academic community. More abstract ontology for mapping people and information on web is FOAF [35]. For understanding teleology relations, we consider a few classes based on the commonest ontology FOAF. We can concentrate on some of the core academic concepts like Person, Document and Group and relationships between them. Fig. 1 presents FOAF relations: (1) publications (domain Person, range Document), where publications are linked to the person, (2) member (domain Group, range Agent) where the Agent as Person can be a member of the Group. The shortcomings of folksonomies mentioned in section 2.3 may be effectively reduced by a collaborative tagging system, that allows the creation of tags in proposed ontology classes.

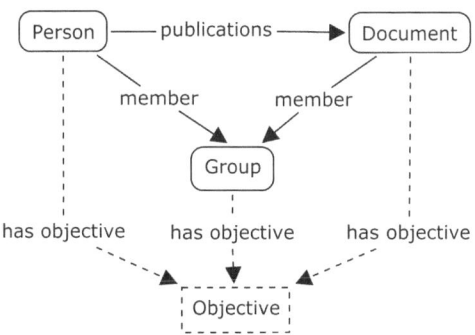

Fig. 1. FOAF ontology classes extend by Objective class

If we consider further the involvement of the Objective, the resulting folksonomy content can bring more benefits for its users. Suggested system helps the participants to formulate their research objectives. Objectives are used as a pointer to the strategic knowledge. Thus, associations between strategic knowledge and other key factors (documents, groups, etc.) can be mapped. Principles of ontology Friend-Of-A-Friend can be used in many different contexts. Often it can be found in social networks such as Facebook, Twitter, Google+, last.fm and others. However, the common element of these networks is that pointing out an objective to be reached is not needed. People share everyday statuses, entertaining content, music or videos they like, pictures and everything gets tagged, liked, re-shared and commented. None of it, though, follows neither middle-term nor long-term objectives. The situation is entirely different in academic or business environment. Whether it is a business objective (e.g. maximizing profit, turnover, market share, etc.) or research objective (developing new technology, getting a patent, publishing search results) progress in these areas are always connected with strategic knowledge. Connecting objectives as a strategic knowledge dimension to FOAF ontology is therefore desired and essential requirement in order to watch and evaluate any progress in chosen strategy.

It is the authors' objective to develop a social network to share research objectives within academic environment that would enhance cooperation between university's ongoing project teams, excganging experts from diverse areas and provide a guidepost for academics as well as students. Teleology serves as a bridge between FOAF ontology and extended ontology with objectives, thus creating a strategic knowledge map. Furthermore, such a network with certain adjustments should be generic enough for deployment in business environment as well. The additional semantic information identifies particular types of knowledge, e.g. we can consider a simple ontology shown on Fig. 1:

- **Person** - tag of type Person is identification of user, e.g. a company worker (Knowing who), who is owner of procedural knowledge (Knowing how).
- **Document** - tag of type Document specifies content as an information resource, declarative knowledge can be included in such document (Knowing what).
- **Objective** - tag of type Objective serves to locate strategic knowledge (Knowing why).

Understanding of a domain and knowledge localization is improved by adding semantic to tags. According to tag type, all shared content (documents, presentations, videos, blog posts, and others) can be filtered. Strategic knowledge map is created by adding objective tag type and its filtering. Such online system functions help people to use other view of domain, based on filtering tag types, when they are locating company knowledge sources. Teleology helps to identify the objective as an essential and core element of the system that points to strategic knowledge. It is an expected path leading to desired results, it is a reason for all steps taken to achieve them, it is a tool to know more about ourselves. By sharing our objectives and associating them with other mapped topics in a knowledge map we trace our work and give others the information how results have been achieved.

4   Conclusion

This paper provided an insight into principles of teleology perspective of knowledge map. Knowledge mapping can be understood in two ways. There is a theory behind division knowledge into different knowledge types and processes of how to create, sort or share knowledge. But it also can be viewed as a knowledge mapping tool developed in appropriate technology that provides support of these features for users.

Essential part of sorting the content in this way is collaborative tagging. When users assign tags to content they work with, they distinguish important resources, that can be then easily navigated to, from the unimportant ones. It is a simple and effective way how to overcome known shortcomings of knowledge mapping and it has several positive impacts on the community. We discussed a possible way of creating and sharing strategic knowledge in academic environment via defining each user's objectives. Proposed ontology model helps to find who is working on what and helps knowledge map users to concentrate on topics of their interest and focus on work towards objectives. An objective is a teleological class in ontology used for sorting content in a knowledge mapping repository. The other classes help to point us to descriptive or procedural knowledge, but objectives provide teleological purpose for actions and other outputs shared within the map.

Acknowledgments. This article was supported by the project INDOP No. CZ.1.07/2.2.00/28.0327, financed from EU and Czech Republic funds.

References

1. Collins, H.: Tacit and Explicit Knowledge. University Of Chicago Press (2010)
2. Nickols, F.: The Knowledge in Knowledge Management,
   http://www.nickols.us/knowledge_in_KM.pdf
3. Holsapple, C.: Handbook on Knowledge Management 1:Knowledge Matters. Springer (2004)
4. Solaz-Portols, J.J., Lpez, V.S.: Types of knowledge and their relations to problem solving in science: Directions for practice. Educational Sciences Journal 06, 105+ (2008)
5. Sveiby, K.E.: The New Organizational Wealth: Managing and Measuring Knowledge-Based Assets. Berrett-Koehler Publishers (1997)
6. Davenport, T.H., Prusak, L.: Working Knowledge. Harvard Business Review Press (2000)
7. Anderson, J.R., Bower, G.H.: Human Associative Memory (Experimental psychology series). Psychology Press (1980)
8. Hamdan, A.M., Saiyd, N.A.: A framework for expert knowledge acquisition. International Journal of Computer Science and Network Security 10, 145–151 (2010)
9. Huff, A.S., Jenkins, M.D.: Mapping Strategic Knowledge. Sage Publications Ltd. (2002)
10. Tandukar, D.: Knowledge Mapping. Ezine Articles,
    http://ezinearticles.com/?Knowledge-Mapping&id=9077

11. Eppler, M.J.: A Process-Based classification of knowledge maps and application example. Knowledge and Process Management 15, 59–71 (2008)
12. Topic Maps Data Model, http://www.isotopicmaps.org/sam/sam-model/
13. Yeung, C.A., Gibbins, N., Shadbolt, N.: Collective user behaviour and tag contextualisation in folksonomies. In: Web Intelligence and Intelligent Agent Technology, pp. 659–662. IEEE (2008)
14. Gordon-Murnane, L.: Social bookmarking, folksonomies, and web 2.0 tools. Searcher Mag. Database Prof. 14, 26–38 (2006)
15. Vaishar, A.: Folksonomie. Inflow: Information Journal 1 (2008)
16. Liu, L., Li, J., Lv, C.: A method for enterprise knowledge map construction based on social classification. Syst. Res. 26, 143–153 (2009)
17. Schifanella, R., Barrat, A., Cattuto, C., Markines, B., Menczer, F.: Folks in folksonomies: social link prediction from shared metadata. In: Proceedings of the Third ACM International Conference on Web Search and Data Mining, pp. 271–280. ACM, New York (2010)
18. Oracle WebCenter, http://www.oracle.com/technetwork/middleware/webcenter
19. IBM Connections,
    http://www-01.ibm.com/software/lotus/products/connections/
20. Angeletou, S.: Semantic enrichment of folksonomy tagspaces. In: Sheth, A.P., Staab, S., Dean, M., Paolucci, M., Maynard, D., Finin, T., Thirunarayan, K. (eds.) ISWC 2008. LNCS, vol. 5318, pp. 889–894. Springer, Heidelberg (2008)
21. Gruber, T.R.: Toward principles for the design of ontologies used for knowledge sharing. Int. J. Hum. Comput. Stud. 43, 907–928 (1995)
22. Nelson, K.Y.L., Seung, H.K., et al.: Ontology-based Collaborative Interorganizational Knowledge Management Network. Interdisciplinary Journal of Information, Knowledge, and Management 4 (2009)
23. Mathes, A.: Folksonomies - cooperative classification and communication through shared metadata. Computer Mediated Communication (2004)
24. Kim, H.-L., Breslin, J.G., Yang, S.-K., Kim, H.-G.: Social Semantic Cloud of Tag: Semantic Model for Social Tagging. In: Nguyen, N.T., Jo, G.-S., Howlett, R.J., Jain, L.C. (eds.) KES-AMSTA 2008. LNCS (LNAI), vol. 4953, pp. 83–92. Springer, Heidelberg (2008)
25. Echarte, F., et al.: Self-adaptation of ontologies to folksonomies in semantic web. Engineering and Technology 33, 335–341 (2008)
26. Schwartz, D.G.: An Aristotelian View of Knowledge for Knowledge Management, pp. 39–48. IGI Global (2011)
27. Encyclopedia.com, http://www.encyclopedia.com/topic/teleology.aspx
28. Rosenberg, A., Arp, R.: Philosophy of Biology: An Anthology. Wiley-Blackwell (2009)
29. Nonaka, I., Takeuchi, H.: The Knowledge-Creating Company: How Japanese Companies Create the Dynamics of Innovation. Oxford University Press (1995)
30. Sweeney, J.D.: A Teleological Approach to Robot Programming by Demonstration. Open Access Dissertations (2011)
31. Judelman, G.: Knowledge Visualization: Problems and Principles for Mapping the Knowledge Space. International School of New Media (2004)
32. Pfeffer, J., Sutton, R.I.: The Knowing-Doing Gap: How Smart Companies Turn Knowledge into Action. Harvard Business School Press (2000)
33. AKT Reference Ontology, http://www.daml.org/ontologies/322
34. Academic Institution Internal Structure Ontology (AIISO),
    http://vocab.org/aiiso/schema
35. FOAF Vocabulary Specification, http://xmlns.com/foaf/spec/

Boosting Retrieval of Digital Spoken Content

Bernardo Pereira Nunes[1,2], Alexander Mera[1],
Marco A. Casanova[1], and Ricardo Kawase[2]

[1] Department of Informatics - PUC-Rio - Rio de Janeiro, RJ - Brazil
{bnunes,acaraballo,casanova}@inf.puc-rio.br
[2] L3S Research Center, Leibniz University Hannover, Germany
{nunes,kawase}@l3s.de

Abstract. Every day, the Internet expands as millions of new multimedia objects
are uploaded in the form of audio, video and images. While traditional text-based
content is indexed by search engines, this indexing cannot be applied to audio and
video objects, resulting in a plethora of multimedia content that is inaccessible to
a majority of online users. To address this issue, we introduce a technique of automatic, semantically enhanced, description generation for multimedia content.
The objective is to facilitate indexing and retrieval of the objects with the help of
traditional search engines. Essentially, the technique generates static Web pages
automatically, which describe the content of the digital audio and video objects.
These descriptions are then organized in such a way as to facilitate locating corresponding audio and video segments. The technique employs a combination of
Web services and concurrently provides description translation and semantic enhancement. Thorough analysis of the click-data, comparing accesses to the digital
content before and after automatic description generation, suggests a significant
increase in the number of retrieval items. This outcome, however is not limited
to the terms of visibility, but in supporting multilingual access, additionally de-
creases the number of language barriers.

Keywords: publishing multimedia content, spoken content retrieval, spoken lecture processing.

1 Introduction

The Internet has veritably become the predominant source of information. What began
as mere textual information within simple hypertext systems has evolved mutually with
technical hardware and broadband Internet access and shifted the simple Web concept
to one of a complex multimedia system. Where previously, the exchange of information
on the Web was chiefly one-way, i.e., webmasters published content to general users,
the catch up of Web 2.0 has opened up new means of interaction in which ordinary
Internet users were given the tools to contribute with their own content. As a result of
these two factors, together with the dissemination of electronic devices with built-in
digital cameras and audio recorders, multimedia content has proliferated tremendously
and become an important source of information, communication and social interaction
on the Web. The product is the emanation of a multimedia phenomenon that marks the
online content we see today. Millions of new images, videos and audio are uploaded

M. Graña et al. (Eds.): KES 2012, LNAI 7828, pp. 153–162, 2013.
© Springer-Verlag Berlin Heidelberg 2013

to the web on a daily basis, spurring and motivating an expanse of new research in various fields. Our focus lies with the issues involving search, retrieval and access to multimedia content, specifically that of audio and video objects, which we refer to as "spoken content".

Today, search engines are the gatekeepers to information. Almost all information accesses begin with a keyword search. Spoken content, unlike its text-based rival, still cannot be indexed by search engines as it is encoded into digital audio and video objects which do not contain intrinsic textual description. Retrieval of information of this type from a keyword search engine is therefore based solely on the few existing metadata (title, description, author, etc), which is of low quality and descriptiveness. As an alternative, content-independent metadata has been acceptably employed to describe multimedia files over the last decade [3,9].

Multilingual accessibility, however has posed another problem to online content discovery. While automatic translation tools do a rather good job translating textual documents and websites, very few have been proven to support the cross-language re-trieval of objects. To confront the issue of multimedia access on the Web caused by the lack of indexable contextual content and language barriers, in this work we present a mash-up tool that facilitates the indexing and retrieval of spoken content through the automatic generation of transcripts, semantic annotations and translation.

The contributions of this work are twofold:

– We provide an online tool which automatically generates semantically enriched transcripts and translations of spoken content.
– An evaluation of the technique as pertains to real learning objects (spoken content).

In accordance with the outcomes provided by our contributions, we aim to answer the following research questions:

– To what extent can automatic generated scripts improve the retrieval of spoken content?
– To what extent do automatic translations of scripts enable the use of spoken material?

The remainder of the paper is structured as follows. Section 2 describes relevant information from previous work. Section 3 introduces our publishing technique. Section 4 exposes the results of implementing our tool. Finally, Section 5 presents discussions, conclusions and future works.

2   Related Work

A recent study explored the improvement of video retrieval via automatic generation of tagging and geotagging [8]. This work concurs with similar approaches which most audio, video and image repositories [2,15,17] on the Web utilize to index multimedia les. However, while content-independent metadata, such as title or author, can describe some aspects of spoken content, the actual content of these objects remains inaccessible to text-based search engines. This results in a tedious search on behalf of the user, because even if the spoken content can be found, the user must still manually locate the specific segment he is seeking. In order to increase the likelihood of retrieving the time-aligned segment of spoken content rather than just the file, a more elaborate form of annotation is required. This content-descriptive form of metadata, which transcribes audio and video content, is, however, a wary task rarely executed by the publisher and/or creator of the content.

Alberti et al. [1] addresses the spoken content retrieval problem in the context of last US presidential campaign, where a scalable system that makes the video content searchable was developed. Although their approach adhibits content-descriptive metadata and content-independent metadata to describe spoken content, the content is not prepared to be machine-readable. To address this issue and present a machine-readable approach, Repp et al. [14] apply a specific ontology to annotate content and make it attainable using OWL-DL for semantic search engines. This approach provides semantic information to search engines, however that same semantic information is not available to assist human Web users. A more proactive approach would be the adoption of RDFa [16], which would assist both machines and humans to index and retrieve Web content [6].

Glass et al. [5] discuss making spoken lectures findable by implementing components of a spoken content retrieval system. Yet this approach poses a challenge to the data management community [9,10,7] because transcription files can only be accessed through a search form and are therefore still hidden from search engines. To pragmatically combat these issues we present a technique of automatic video and audio text description generation, which transforms the spoken search problem into a traditional text search problem, and in so doing makes spoken content available to users on the Web. Additionally, our technique processes spoken content in a manner which permits text-based search engines to assist in locating time-aligned segments of the content.

Furthermore, the technique also annotates entities [11] present in the transcriptions using RDFa. The annotation process employs Dublin Core [4] to explicate the content-independent metadata of an asset, while the content-descriptive metadata applies a specic set of ontologies from DBpedia[1] to illustrate the concepts and relationships with other Web resources. Consequently, content provenance is known by search engines once that content has been linked with other resources, thus improving precision page ranking.

The effectiveness of the proposed technique is described by an experiment with over a thousand minutes of spoken content, divided into 99 video objects and an in-depth hit analysis of these objects.

3   Publishing Technique

This section describes the publishing process with the aid of a complete spoken content publishing example (see Figure 1). Using an automatic speech recognition service (ASR), the first step is to transcribe a given spoken content [12,13]. The set of time-aligned text excerpts thus obtained forms what we call the script of the spoken content, by analogy of the usual meaning of the word. This approach thus converts the spoken search problem to a text search problem since the script is a textual representation of the spoken content. Figure 2 shows an example of a script.

---

[1] http://www.dbpedia.org

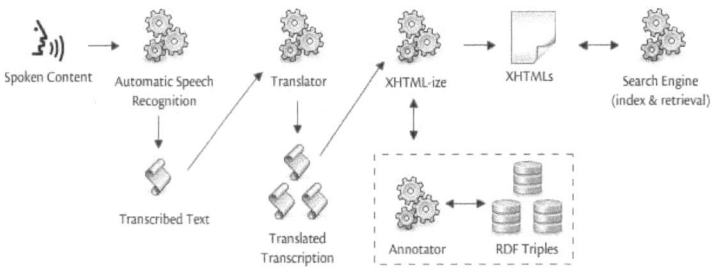

Fig. 1. Publishing technique

Step two seeks to reach additional user populations by translating, if desired, the script into other languages. Our technique translates these scripts into various languages using the Google Translator API[2]. In the following section, we discuss how this step impacts content retrieval.

The final step is decomposed into two substeps: (a) to transform a plain text script into a XHTML (eXtensible Hypertext Markup Language) file; and (b) to annotate the content using RDFa (Resource Description Framework in attributes).

```
<!DOCTYPE HTML PUBLIC "-//W3C//DTD HTML 4.01//EN" "http://www.w3.org/TR/html4/strict.dtd">
<html xmlns="http://www.w3.org/1999/xhtml" xml:lang="en">
...
<div>
   <a href="URL?time=04m47s">Watch this excerpt</a>
   <p>
       Against this backdrop comes Fritz Haber. The man
       who carried out the synthesis of ammonia in order
       to produce it on an industrial scale, from
       molecular hydrogen and nitrogen, abundantly
       available.
   </p>
</div>
```

Fig. 2. Example of a script

The first substep involves a conversion process, in which static Web pages for each asset are generated and every time-aligned excerpt of text from a spoken content is recast into sections (div elements). Each of these sections contains a hyperlink (the a element) that indicates the exact segment of spoken content where the speech occurs and the transcribed text related to that content (p element) (see Figure 2). Moreover, the language and document type of the content is specified by each static Web page generated by the aforementioned method. In the example of Figure 3, the document type is XHTML (4.01 strict) and the language of the content is in English ("xml:lang=en").

The code in Figure 2 is not enriched with semantic markups, although XHTML supports semantic description. Accordingly, the embedding of a collection of attributes into XHTML markups to enrich the semantics of the Web page content comprises substep

---

[2] http://code.google.com/apis/language/

```
  . . .
      Against this backdrop comes <a
      about="http://dbpedia.org/resource/Fritz_Haber"
      typeof="http://dbpedia.org/ontology/Scientist"
      href="http://dbpedia.org/resource/Fritz_Haber"
      title="http://dbpedia.org/resource/Fritz_Haber">F
      ritz Haber</a>.

  . . .
```

Fig. 3. Example of a Web page

two. Spotlight Web Services analyzes the content transcribed in the previous step, by annotating references to DBpedia resources in the text. Thus, the text is enriched with entity detection and name resolution by using. Figure 4 shows the annotation result. The potential exists in this substep to provide a solution for linking the Linked Open Data (LOD) cloud to unstructured information sources via DBpedia. Thus, the annotated text can be used for secondary tasks, such as recommending related assets based on semantics and displaying additional information about those assets in congruence with its primary use to enhance search and retrieval.

Against this backdrop comes Fritz Haber. The man who carried out the synthesis of ammonia in order to produce it on an industrial scale, from molecular hydrogen and nitrogen, abundantly available.   http://dbpedia.org/resource/Fritz_Haber

Fig. 4. Result of the XHTML-ize step

4   Experiments

Over the course of our study, real Web data from the educational domain was used to perform an extensive evaluation of our spoken content publishing technique. Our objectives included both a thorough analysis of Web page hits synthesized for the audio and video learning objects, as well as an assessment of the efficacy of the publishing technique.

4.1   Experimental Setup

The tool was evaluated by means of 99 Learning Objects (LO's) comprised of 10 minute video files each containing dialogs on diverse elementary chemistry topics. The experiment was carried out over a period of eight months in two separate stages. Stage one lasted five months and during this time all 99 LO's were published using content-independent metadata. The objective of the first stage was to equalize the Web page hits via content-independent metadata. The information gathered was then used to create two balanced groups of LO's to evaluate the efficacy of the tool. Each LO was then sorted according to the number of hits and further assigned to different groups in pairs who shared the same order of magnitude.

The second stage was a selective process in which one of the groups was submitted to the tool, while the other was not. This stage had a duration of three months. The objective of this second stage was to evaluate the efficacy of the tool. Static Web pages were hosted on the Wordpress server, while the video files were hosted on Youtube. The tool was then assessed using the statistics these services provide.

4.2   Data Analysis

Let Group A refer to the set of LO's published in the first stage, Group P refer to the set of LOs published using the tool in the second stage, and Group $\neg P$ refer to the set of LOs described only by content-independent metadata. This section examines the results of the hit analysis during both experiment stages.

Total Hits Analysis. Approximately 75K hits were obtained during both stages of the experiment. Group A, which represents the first stage with all 99 LO's, obtained just 22% of the total number of hits. It is important to note that in stage one, data was collected for five months, whereas stage two had a three-month duration, thus 78% of the hits were performed in stage two (see Figure 5).

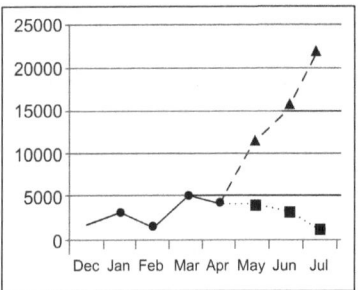

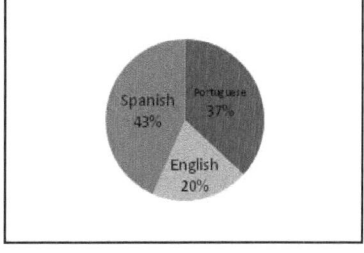

Fig. 5. Hit stats. First stage 1-5. Second stage 5-8

Fig. 6. Hits percentage of translated static Web pages generated by our publishing technique

Observe that Group P captured 66% of the total number of hits whereas Group $\neg P$ just 12%. Hence, Group P captured 84% of the total number of hits in the second stage, i.e., 5.3 times more than the number of hits obtained by Group $\neg P$.

Page Hits Analysis. As described in Section 3, a new static Web page was created for each new language in the translation step. Throughout the experiment, three language scripts were generated for each asset; English, Spanish and Portuguese. Figure 6 provides the percentage of total number of hits for each translated static Web page: 43% for pages in Spanish, 37% for pages in Portuguese and 20% for pages in English.

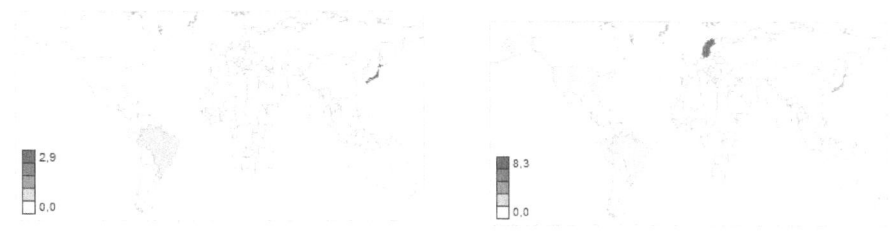

Fig. 7. Countries that have interacted with the content, on the left: Japan, Brazil and Portugal. On the right: Sweden, Japan, Brazil, Spain, Peru, USA and Portugal.

Although static Web page visitors are anonymous, that is, no information pertaining to his/her location or mother language is provided, this information is available for logged in users on Youtube.

The information attained from Youtube is highly relevant to our study because all actions (share, comment or mark as favorite) are executed by users that share in interest in the content of an asset. According to this information and as depicted in the left image of Figure 7, only users from Brazil, Portugal and Japan shared, commented or marked as favorites the assets of Group A.

The right image of Figure 7 reveals that after the technique had been applied creating Group P, users from other countries (Sweden, Japan, Brazil, Spain, Peru, United States and Portugal) could be reached. Note that, the native language of Brazil, Spain, Peru, United States and Portugal is indeed English, Spanish or Portuguese, the languages implemented during experimentation with our technique. This does not extend, however to the populations of Japan and Sweden, although there is a sizable population of Brazilians living in Japan.

Table 1. Bottom 10 most accessed Learning Objects

LO's with the largest number of hits	1	2	3	4	5	6	7	8	9	10
Group $\neg P$	1636	1615	1416	969	812	642	514	300	265	220
Group P	2466	1774	1744	1562	1499	1467	1421	1414	1386	1307

Assets Hits Analysis. At this stage of the experiment, the variance between the number of hits of an LO that was published using the tool and an LO described by content-independent metadata was addressed. Table 1 depicts the top 10 LO's, according to the number of hits, where the second line corresponds to LO's in Group P and the third line to LO's in Group $\neg P$. Table 1 illustrates, as expected, that the number of hits for LO's indexed by the tool (Group P) is notably greater than the number of hits for LO's described only by content-independent metadata (Group $\neg P$). This is generally true for all 99 LO's. Table 2 depicts the LO's with the lowest number of hits, and shows an even greater discrepancy.

Table 2. Top 10 most accessed Learning Objects

Least hit LO's	10	9	8	7	6	5	4	3	2	1
Group ¬P	32	27	27	25	24	21	21	12	11	7
Group P	333	326	287	271	250	213	195	191	108	95

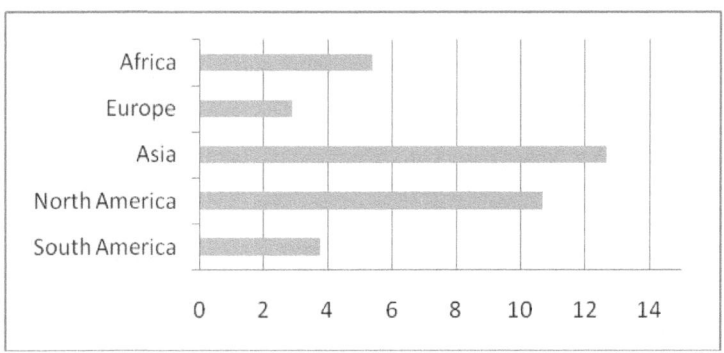

Fig. 8. The number of hits boosted in different orders of magnitude for each continent

Regional Analysis. Using the information gathered from Youtube pertaining to user location, it was possible to tabulate the number of hits by continent (with the exception of Oceania). In the stage one of analysis, Group A obtained 13,911 hits from South America, 73 hits from North America, 39 hits from Asia, 1,297 hits from Europe and 31 hits from Africa. We note that, from the hits in South America, 13,673 (approx. 99%) were from Brazil, a Portuguese-speaking country. The same was observed in Europe where, out of the 1,297 hits, 1,107 (approx. 85%) were from Portugal, which is also a Portuguese-speaking country.

In the second stage, the assets captured 52,366 hits from South America, 3,738 hits from Europe, 779 hits from North America, 494 hits from Asia, and 167 hits from Africa. We again highlight that the vast majority of hits from South America came from Brazil, and note that the population of Brazil comprises almost 50% of that of South America. It is important to note, however that during a brief period the number of hits from other South American countries increased from 1% to 5.5%. This increase largely took place in Spanish speaking countries. Similarly, the number of hits obtained from European users was also less concentrated: Portugal, which captured 85% of the hits in the first stage of the experiment, had 67% in the second stage, whereas the total number of hits from other European countries more than doubled from 15% to 33%. Figure 8 depicts the ratio increase by continent, obtained by dividing the number of hits in the second stage of the experiment by the number of hits in the first stage.

5   Conclusion

In this paper we presented a technique which automatically enhances spoken content on the Web using semantic descriptions (transcripts) and translations. This technique facilitates indexing and retrieval of the objects with the aid of traditional text search engines. Our techniques provided us the basis for an online tool which was used to complete the evaluations demonstrated in this paper.

The tool functions by automatically generating static Web pages which describe the spoken content, which are organized to facilitate locating segments of the content corresponding to the descriptions. The tool further annotates the described spoken content using RDFa and DBpedia to link unstructured information sources to the LOD cloud and enhances search and information retrieval of the assets. The tool also provides a means of amplifying user range by breaking down language barriers through the creation of a multilingual resource. Evaluation proves that the number of hits to the objects processed by the tool was significantly improved, as well the access of consumers of foreign languages.

Future work will investigate several extensions to the tool. First, the tool may resort to semantic information to display complementary information about an asset. Second, the semantics will be enriched to function not only with DBpedia resources, but also with other LOD data sources. Finally, we will recommend related assets by taking advantage of the connected text through ontologies from the LOD and to assess its effectiveness.

A complete description of the tool may be found at http://moodle.ccead.puc-rio.br/spokenContent/.

Acknowledgement. This work has been partially supported by CAPES (Process n° 9404-11-2). Additional thanks to Chelsea Candra Schmid for her cooperation.

References

1. Alberti, C., Bacchiani, M., Bezman, A., Chelba, C., Drofa, A., Liao, H., Moreno, P., Power, T., Sahuguet, A., Shugrina, M., Siohan, O.: An audio indexing system for election video material. In: Proceedings of the 2009 IEEE International Conference on Acoustics, Speech and Signal Processing, ICASSP 2009, pp. 4873–4876. IEEE Computer Society, Washington, DC (2009)
2. Baidu search engine, http://www.baidu.com
3. Brezeale, D., Cook, D.: Automatic video classification: A survey of the literature. IEEE Transactions on Systems, Man, and Cybernetics, Part C: Applications and Reviews 38(3), 416–430 (2008)
4. Dublin core metadata initiative, http://www.dublincore.org
5. Glass, J., Hazen, T.J., Cyphers, S., Malioutov, I., Huynh, D., Barzilay, R.: Recent Progress in the MIT Spoken Lecture Processing Project. In: Proc. Interspeech (2007)
6. Haslhofer, B., Momeni, E., Gay, M., Simon, R.: Augmenting europeana content with linked data resources. In: Proceedings of the 6th International Conference on Semantic Systems, I-SEMANTICS 2010, pp. 40:1–40:3. ACM, New York (2010)

7. Jiang, L., Wu, Z., Zheng, Q., Liu, J.: Learning deep web crawling with diverse features. In: Proceedings of the 2009 IEEE/WIC/ACM International Joint Conference on Web Intelligence and Intelligent Agent Technology, WI-IAT 2009, vol. 01, pp. 572–575. IEEE Computer Society, Washington, DC (2009)

8. Larson, M., Soleymani, M., Serdyukov, P., Rudinac, S., Wartena, C., Murdock, V., Friedland, G., Ordelman, R., Jones, G.J.F.: Automatic tagging and geotagging in video collections and communities. In: Proceedings of the 1st ACM International Conference on Multimedia Retrieval, ICMR 2011, pp. 51:1–51:8. ACM, New York (2011)

9. Madhavan, J., Afanasiev, L., Antova, L., Halevy, A.: Harnessing the Deep Web: Present and Future. In: 4th Biennial Conference on Innovative Data Systems Research (CIDR) (January 2009)

10. Madhavan, J., Ko, D., Kot, L., Ganapathy, V., Rasmussen, A., Halevy, A.: Google's deep web crawl. Proc. VLDB Endow. 1, 1241–1252 (2008)

11. Mendes, P.N., Jakob, M., García-Silva, A., Bizer, C.: Dbpedia spotlight: shedding light on the web of documents. In: Ghidini, C., Ngomo, A.-C.N., Lindstaedt, S.N., Pellegrini, T. (eds.) I-SEMANTICS. ACM International Conference Proceeding Series, pp. 1–8. ACM (2011)

12. Nexiwave – speech indexing, http://www.nexiwave.com

13. Nuance – dragon naturallyspeaking, http://www.nuance.com

14. Repp, S., Meinel, C.: Automatic extraction of semantic descriptions from the lecturer's speech. In: IEEE International Conference on Semantic Computing, ICSC 2009, pp. 513–520 (September 2009)

15. Truveo video search, http://www.truveo.com

16. W3c – rdfa primer, http://www.w3.org/TR/xhtml-rdfa-primer

17. Youtube – broadcast yourself, http://www.youtube.com

Constructing the Integral OLAP-Model
for Scientific Activities Based on FCA

Tatiana Penkova and Anna Korobko

Institute of Computational Modeling of the Siberian Branch of
the Russian Academy of Science, Krasnoyarsk, Russia
{penkova_t,lynx,}@icm.krasn.ru

Abstract. In this paper an original approach to analytical decision making support based on on-line analytical processing of multidimensional data is suggested. According to Dr. Codd's rules, the effectiveness of data analysis significantly depends on the data accessibility and transparency of an analytical model of domain. The method of constructing a conceptual OLAP-model as an integral analytical model of the domain is proposed. The method is illustrated by the example of the scientific activities domain. The integral analytical model includes all possible combinations of analyzed objects and gives them the opportunity to be manipulated ad-hoc. The suggested method consists in a formal concept analysis of measures and dimensions based on an expert knowledge about the structure of analyzing objects and their comparability. As a result, conceptual OLAP-model is represented as a concept lattice of multidimensional cubes. Concept lattice features allow the decision maker to discover the non-standard analytical dependencies on the set of all actual measures and dimensions of the scientific activities domain. Conceptual OLAP-model implementation allows user makes better decisions based on on-line analytical processing of the scientific activity indicators.

Keywords: Integral OLAP-model, On-line analytical processing, Formal concept analysis.

1   Introduction

The effectiveness of administrative resources management depends on the way the analytical information to be provided. The on-line analytical processing (OLAP) is widely used for analytical decision making support [1-5]. The analytical processing of large amount of data in government (e.g. territorial, industrial, corporate) requires new approaches to be developed for OLAP implementation. The effectiveness of data analysis depends largely on the data accessibility and transparency of an analytical model of domain. As usual, the analytical model of domain is a set of OLAP-models for solving the particular problems [1-3]. Such a situation can be represented as a fragmentary analytical model of the domain. Constructing an integral analytical model is a topical problem in computer science [4-8]. In the papers [4, 6] the integral approach implementation as a constructing of analyzing objects catalogues is – represented. This way allows the engineer to systematize the analyzing objects, but

M. Graña et al. (Eds.): KES 2012, LNAI 7828, pp. 163–170, 2013.
© Springer-Verlag Berlin Heidelberg 2013

doesn't enable the end-user to intuitively manipulate them. Authors [7] suggest method of constructing the integral model based on ontology. This method gives the user to manipulate analyzing objects of ontology, but doesn't provide generating analytical queries. Our papers [8, 9] present the first informal approach to using the formal concept analysis for the on-line analytical processing. To develop the formal method of constructing integral OLAP-model based on an expert knowledge about analyzed objects structure and relation of their comparability is of importance.

In the paper the method of constructing the conceptual OLAP-model as the integral analytical model of the domain is proposed. The suggested method is based on formal concept analysis of domain measures and dimensions. The way of constructing the integral analytical model as a concept lattice of multidimensional cubes is described formally. The integral analytical model includes all the possible combinations of analyzing objects and makes possible the manipulation of them ad-hoc. Conceptual OLAP-model implementation improves the effectiveness of decision making support based on on-line analytical processing of multidimensional data.

The outline of our paper is as follows. Section II presents the formal description of multidimensional data model and shows importance of developing the integral OLAP-model. Section III describes the method of constructing conceptual OLAP-model using formal concept analysis. In Section IV, we consider the implementation of the method of conceptual OLAP-model constructing for scientific activities of the organization. We conclude in Section V and sketch some issues for future research.

2   On-Line Analytical Processing

The term OLAP (On-line Analytical Processing) is introduced in 1993 by E. Codd [10]. Codd formulated 12 rules of OLAP technology. The realization of these rules depends on current level of information technology development. OLAP provides an efficient means to analyze data. OLAP-means represent data as easy-to-understand and easy-to-use data model, which consists of multidimensional cubes.

The OLAP-cube can be represented as a pair:
$G = <D, F>$, where
$D = <d_1, d_2, ..., d_n>$ is a set of the cube dimensions and $F = <f_1, f_2, ..., f_m>$ is a set of the cube measures (facts). The measure is a numerical characteristic of the analyzed process. And the dimension is an array of values, which belongs to the one data type and characterizes a structural property of the domain. A set of dimensions forms an axis of the cube. There are analytical measures in a cube cell.

OLAP operations (e.g. slicing, dicing, drilling, pivoting, filtering) enable users to navigate data flexibly, define relevant data sets, analyze data at different granularity and visualize results in different forms. An significant advantage of OLAP is an operation with domain terms [11].

The quantity and content of domain OLAP-cubes depends on the quantity of solving particular problems. To solve a new analytical problem, user has to construct a new OLAP-cube. Constructing OLAP-cube involves the selection of necessary database tables, join of these tables, selection of the table fields and their comparison with domain terms. It requires special knowledge about database scheme, analyzing objects and their properties, about algorithms of analytical measures calculation and about relationships between objects. Thus, to solve each problem the decision maker has to apply to an engineer for assistance. As a result, a set of OLAP-cubes forms a fragmental analytical model of domain.

The analytical model of domain has to include all the actual measures and dimensions and all their combinations to improve the effectiveness of decision making support based on on-line analytical processing. To manipulate all the available analyzing objects the analytical model of the domain should be relied on an expert knowledge about analyzing objects structure and their comparability.

3   The Method of Constructing Conceptual OLAP-Model

To construct the integral OLAP-model based on all the actual analyzing objects, we need to determine groups of objects (measures and dimensions) can be processed together. In this case, it is preferable to use the methods of binary clustering [12]. One of the most appropriate method is data analysis based on formal concepts and concept lattices [13].

Formal Concept Analysis (FCA) is introduced in 1981 by R. Wille [14]. The method is based on understanding the world in terms of objects and attributes. The formal context is a triplet $K = (G, M, I)$, which consists of set G, set M and relation $I \subseteq G \times M$. The elements of set G are objects of the context; the elements of set M are attributes of the context. A binary relation I between G and M (described by gIm) indicates when object $g \in G$, has attribute $m \in M$. For set $A \subseteq G$ and for set $B \subseteq M$ it is defined that: $A' = \{m \in M \mid gIm$ for all $g \in A\}$ (all attributes in M shared by the objects of A); $B' = \{g \in G \mid gIm$ for all $m \in B\}$ (all objects in G that have all the attributes of B). The formal concept of the formal context is defined by derivation operators as pair (A, B) with $A \subseteq G$, $B \subseteq M$, $A = B'$, $B = A'$. A is called an extent, and B is called an intent of concept (A, B).

Formal concept analysis of measures and dimensions allows us to construct the integral OLAP-model based on expert knowledge about analyzed objects and their comparability [15]. The method of constructing conceptual OLAP-model based on integration of OLAP and FCA is suggested. Fig. 1 shows the IDEF0 context diagram of constructing the conceptual OLAP-model.

The method of conceptual OLAP-model constructing includes the following basic stages:

1. Determining a set of analyzed objects.
2. Forming a formal context.
3. Generating the formal cube-concepts.
4. Constructing the concept lattice of multidimensional cubes.

At the first stage, an expert determines analytical problems of domain and forms the queries based on interviewing of the end-users and exploration of the reports. Further, the expert extracts special domain terms by analyzing the analytical queries. The determined domain terms form a set of analyzed objects that is used for OLAP-cubes construction. In accordance with the multidimensional data model, objects of formal concept is divided

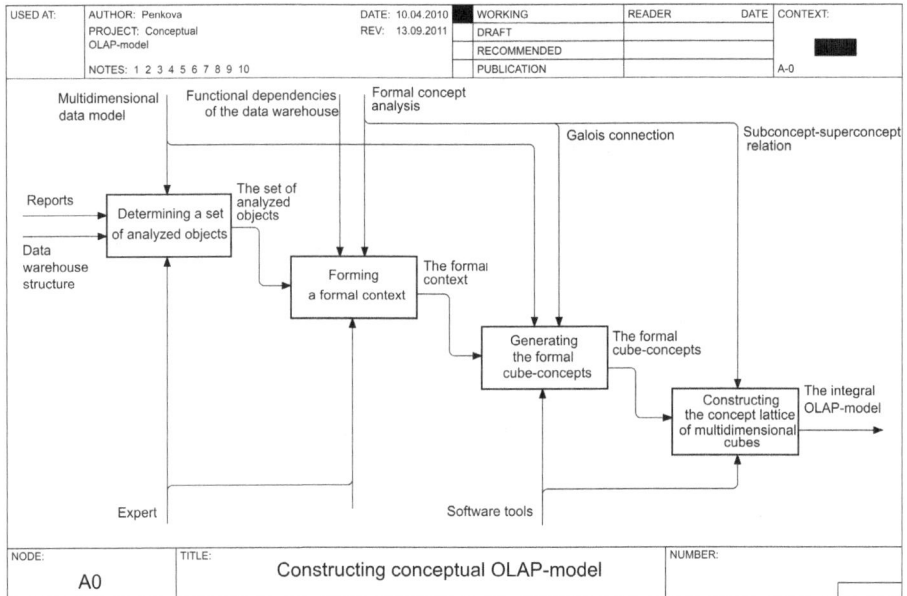

Fig. 1. Constructing conceptual OLAP-model

into two sets: a set of measures (facts) $F = \{f_1, f_2, ..., f_m\}$, and a set of dimensions $D = \{d_1, d_2, ..., d_n\}$. The numeric characteristics are the set of measures, and analysis aspects are the set of dimensions. Then, the expert defines database tables and compares analyzed objects with data fields based on data warehouse structure.

At the second stage, the expert forms a formal context based on knowledge about analyzing objects structure and relation of their comparability. The relation of comparability can be identified as R. In the strict sense, we can say that $(d_i, f_j) \in R$ if the i-th measure can be processed with the j-th dimension. The formal context K in accordance with FCA is defined as a triplet (F, D, R). The elements of set F are measures; the elements of set D are dimensions. The formal context can be represented as a binary matrix where the measures are rows, the dimensions are columns and relation of comparability is a cross at intersection between a row and a column.

At the next stage, the formal cube-concepts are generated based on created formal context. The pair (A, B) is a formal cube-concept, where A is a set of equidimensional measures, which are processed with all dimensions of B. The set of measures A is called the extent, the set of dimensions B is called the intent of the formal concept (A, B). In the terms of OLAP, the formal cube-concept is an analytical multidimensional cube, which is complete with respect to addition of equidimensional measures and compatible dimensions.

At the final stage the concept lattice of multidimensional cubes is constructed. In accordance with FCA the set of all formal concepts of the context is ordered by the subconcept-superconcept relation. For two concepts (A_1, B_1) and (A_2, B_2) this order is formalized as: $(A_1, B_1) \le (A_2, B_2)$: $\Leftrightarrow A_1 \subseteq A_2$ ($\Leftrightarrow B_2 \subseteq B_1$). (A_1, B_1) is called a subconcept of (A_2, B_2), and (A_2, B_2) is called a superconcept of (A_1, B_1). The set of all concepts together with subconcept-superconcept relations forms a complete lattice, which is called a concept lattice [15]. For conceptual OLAP-model, the concept lattice is defined as a concept lattice of multidimensional cubes and the subconcept-superconcept relation is defined as a subcube-supercube relation. It means that measures of a parent cube include measures of a child cube and dimensions of the child cube include dimensions of the parent cube.

The concept lattice of multidimensional cubes is an image of the integral OLAP-model of the domain. The concept lattice features allow us to develop algorithms for intuitively manipulation of all the analyzed objects and analytical experiment support. The integral OLAP-model covers all the possible analytical problems of the domain.

4   The Conceptual OLAP-Model for Scientific Activities

Consider the implementation of the method of conceptual OLAP-model constructing for scientific activities of the organization. Scientific activities are the basic form of activities of any scientific and educational institutes. It is intellectual work aimed at acquisition and deployment of new knowledge for technological, economic, social problem solving.

Estimation of the scientific activities effectiveness is connected with solving the following analytical problems:

- estimation of the scientific research quality;
- estimation of the researcher professionalism;
- estimation of the researcher publish activities;
- estimation of the science and education integration;
- estimation of the teaching practice;
- other problems.

Solving the analytical problems is connected with the following queries:

- published scientific works made by researchers;
- monographs written by researchers;
- papers in international journals published by researchers in a year;
- papers in national journals published by researchers;
- methodical literature published by researchers;
- graduation thesis or papers were defended;
- grants obtained by researchers;
- patents obtained by researchers;
- conferences established by organization;
- honorary titles and prizes got by researchers;
- other queries.

We can determine a set of domain terms by analyzing the queries. Then, we can define the scientific activities analysis objects: a set of measures (e.g. number of published works, number of obtained patents, number of published papers, number of published methodical literature, number of established conferences, number of researchers, etc.) and a set of dimensions (e.g. type of published work, journal name, type of methodical literature, type of patent, author, status of conference, department name, city, year, etc.).

The formal context is formed based on the comparability relation between measures and dimensions according to expert opinion. The relation of comparability is defined if a measure and a dimension can be processed together. Fig. 2 shows the context fragment of scientific activities.

		d_1	d_2	d_3	d_4	d_5	d_6	d_7	d_8	d_9
		Year	Department name	Type of methodical literature	City	Journal name	Type of published work	Type of patent	Status of conference	Author
f_1	Number of published works	✗	✗				✗			✗
f_2	Number of established conferences	✗	✗		✗				✗	
f_3	Number of obtained patents	✗	✗					✗		✗
f_4	Number of published papers	✗	✗		✗	✗	✗			✗
f_5	Number of published methodical literature	✗	✗	✗	✗		✗			✗

Fig. 2. Context fragment of the of scientific activities

The described context is represented by set of measures F = {Number of published works, Number of formed conferences, Number of obtained patents, Number of published articles, Number of published methodical literature} and set of dimensions D = {Year, Department name, Type of methodical literature, City, Journal name, Type of published work, Type of patent, Status of conference, Author}. Using numbers and letters as abbreviations, we also can write F = {f_1, f_2, f_3, f_4, f_5} and D = {d_1, d_2, d_3, d_4, d_5, d_6, d_7, d_8, d_9}. Relation R is represented by the crosses in context can be formally be expressed by R = {(f_1, d_1), (f_1, d_2), (f_1, d_6), (f_1, d_9), ..., (f_5, d_9)}.

The formal cube-concepts are generated based on created context. In our example, the formal context contains 10 formal cube-concepts. Consider the way of formal cube-concept forming in detail using the derivation operators:

Choose any set of the measures: A = {Number of formed conferences; Number of published articles} = {f_2, f_4}.

Derive the compatible dimensions: A' = {year; department name; city} = {d_1, d_2, d_4}.

Derive the measures which can be processed with all derived dimensions: A'' = {Year; Department name; City}' = {Number of formed conferences; Number of published articles; Number of published methodical literature} = {f_2, f_4, f_5}.

The set of the measures A'' = A = {f_2, f_4, f_5} can be processed only with set of the dimensions A' = B = {d_1, d_2, d_4} at the same time. Therefore, the pair (A, B) = {{f_2, f_4, f_5}, {d_1, d_2, d_4}} is a formal cube-concept.

The concept lattice of multidimensional cubes is constructed based on subcube-supercube relation. Fig. 3 shows the integral OLAP-model of the scientific activities as the cub-concept lattice.

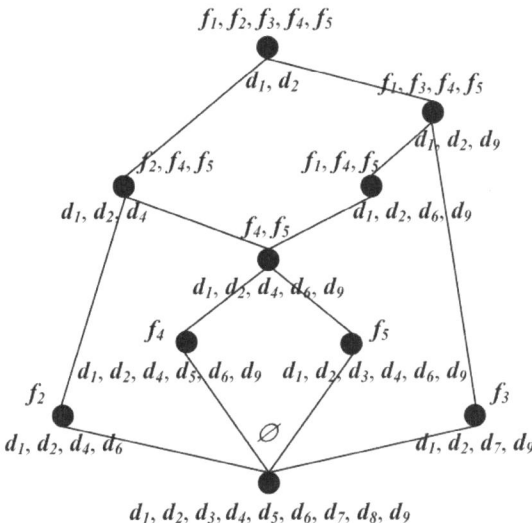

Fig. 3. Cube-concept lattice of the scientific activities

The conceptual OLAP-model includes all the possible combinations of analyzing objects and enables the decision maker to manipulate them. When user selects measures and dimensions, assigned for analysis, the conceptual OLAP-model allows algorithm to determine the set of cubes are in accord with user's query. The set of determined multidimensional cubes allows us to determine the additional measures and dimensions to be processed with selected analyzing objects. The discovery of analytical dependencies of the set of analyzed objects improves the effectiveness of decision making process based on on-line analytical processing.

5   Conclusion

This paper has presented an approach to analytical decision making support based on on-line analytical processing of multidimensional data. The suggested method of constructing conceptual OLAP-model allows us to form the integral analytical model of domain as a lattice of multidimensional cubes. It allows the decision maker to manipulate analyzing objects in accordance with expert knowledge about domain structure and relation of objects comparability. The conceptual OLAP-model includes all the possible combinations of analyzed objects and enables the end-user to discover the nonstandard analytical dependencies on the set of all actual analyzing objects.

The conceptual OLAP-model implementation improves the effectiveness of decision making support based on on-line analytical processing of multidimensional data.

The future research will be connected with developing the algorithm of mining analytical dependencies based on the conceptual OLAP-model; developing the method for forming knowledge base as the concept lattice of multidimensional cubes. Developing the software tools for intelligent analytical decision making support based on suggested approach is expected.

Acknowledgment. This paper was supported by a grant of the Fundamental investigations of the Russian Academy of Science for 2012-2014 (No 15.3) and by a grant of the Russian Fond of Fundamental Researchers for 2012-2013 (No 12-07-31143).

References

1. Gorelov, B.A., Gorelov, B.B.: Model data constructing for on-line analitical processing of the financial information of the University. J. University Management 4(23), 33–46 (2002)
2. Nozhenkova, L.F., Shaidurov, V.V.: OLAP technologies of on-line analytical support of administration. J. Information Technology and Computing Systems 2, 15–27 (2010)
3. Palaniappan, S., Ling, C.: Clinical Decision Support Using OLAP With Data Mining. International Journal of Computer Science and Network Security 8(9), 290–296 (2008)
4. Javed, A., Shaikh, M., Bhatti, B.: Conceptual Model for Decision Support System Based Business Intelligence OLAP Tool for Universities in Context of E-Learning. In: World Congress on Engineering and Computer Science, pp. 826–830 (2008)
5. Vasilecas, O., Smaizys, A.: Business rule based data analysis for decision support and automation. In: International Conference on Computer Systems and Technologies, vol. II (2006)
6. Bojarsky N.A., Shovkun A.V.: Analytical part of information-analytical sysytem constracting by Oracle OLAP Option и BI Beans, Oracle Magazine RE (2004) , http://www.interface.ru/home.asp?artId=9604
7. Lee, J., Mazzoleni, P., Sairamesh, J., Touma, M.: System and method for planning and generating queries for multi-dimensional analysis using domain models and data federation. United States Patent: US 2008/7337170 B2
8. Korobko, A., Penkova, T.: OLAP-modeling of municipal procurement automation support problem. In: International Conference on Conceptual Structures, pp. 87–91 (2009)
9. Korobko, A., Penkova, T.: On-line analytical processing based on Formal concept analysis. J. Procedia Computer Science 1(1), 2305–2311 (2010)
10. Codd, E.F.: Providing OLAP to user-analysts: An IT mandate. Codd and Associates (1993)
11. Gray, J., Bosworth, A., Layman, A., Priahesh, H.: Data Cube: A Relational Aggregation Operator Generalizing Group-By, Cross-Tab, and Sub-Totals. In: 12 th International Conference on Data Engineering, pp. 152–159. IEEE (1995)
12. Kedrov, S., Kuznetsov, S.: Study of Groups of Web Users with Methods Based on Formal Concept Analysis and Data Mining. J. Business-informatics (1), 45–51 (2007)
13. Ganter, B., Wille, R.: Formal Concept Analysis: mathematical Foundations. Springer, Heidelberg (1999)
14. Wille, R.: Restructuring Lattice Theory: an approach based on hierarchies of concept, pp. 445–470. Reidel, Dordrecht (1982)
15. Birkhoff, G.: Lattice theory, 568 p. Nauka, Moscow (1984)

Reasoning about Dialogical Strategies

Magdalena Kacprzak[1] and Katarzyna Budzynska[2]

[1] Faculty of Computer Science, Bialystok University of Technology, Poland,
[2] Institute of Philosophy and Sociology of the Polish Academy of Sciences, Poland

Abstract. The paper proposes an extension of the modal logic $\mathcal{AG}_n$ with operators for reasoning about different types of strategies which agents may adopt in order to win a dialogue game. We model agent communication using the paradigm of formal systems of dialogues and in particular, a system proposed by Prakken. In the paper, the traditional notion of a winning strategy is extended with a notion of a strategy giving a *chance* for success and a notion of a strategy giving a *particular degree* of chances for victory. Then, using the framework of Alternating-time Temporal Logic (ATL) we specify $\mathcal{AG}_n$ operators which allow the investigation of the dialogical strategies.

Keywords: formal systems for dialogues, dialogue games, strategies to win a dialogue, modal logics, modal operators for strategies.

1   Introduction

The paper proposes the specification of modal operators for reasoning about different types of dialogical strategies. The motivation is to provide formal models of dialogue (see e.g. [11,13]) with the possibility of investigating the strategies that agents may adopt during a dialogue game.

Modelling agent communication as a dialogue govern by a specific set of rules attracts a lot attention lately, since such an approach allows not only the representation of speech (communication) acts that agents perform (like e.g. in FIPA-ACL [1]), but also the regulation of their interaction during a dialogue (cf. [10]). Most of dialogue systems provide six types of rules which regulate how a game should be played (cf. [13]): (1) locution rules which describe what type of speech acts players can execute during a dialogue (e.g. an agent i may be allowed to use: *claim φ* for asserting proposition φ, *why φ* for challenging φ, and *retract φ* for withdrawing the commitment to φ), (2) protocol which describes what kind of speech acts a player can execute in all conditions of a dialogue (e.g. after *why φ* an agent i can perform: *claim ($\psi, \psi \rightarrow \varphi$)*, or *retract φ*), (3) effect rules which describe how a particular speech act affects the commitment base of the player (e.g. the performance of *claim φ* by an agent i results in adding φ into i's commitment base), (4) outcome rules which define the outcome of a dialogue (e.g. in persuasion dialogue, i wins when after a dialogue game i's standpoint φ is in the commitment base of its opponent j), (5) turntaking rules which determine a turn of a dialogue, and (6) termination rules which determine the cases where no move is legal.

M. Graña et al. (Eds.): KES 2012, LNAI 7828, pp. 171–184, 2013.

In each dialogue game, an agent has an individual goal of engaging in the message exchange (related to the outcome rules of a dialogue system). For example, in a persuasion dialogue an agent i has the goal of persuading the other party j, i.e. i aims to influence j to adopt i's standpoint [16]. In such a case, the question how to achieve the goal, i.e. what strategy i should apply, becomes crucial (see e.g. [9,14,3] for the dialogue systems that consider the strategies of communicating agents). Typically, formal systems of dialogue uses a game-theoretical notion of a strategy, i.e. a notion of a winning strategy. Such a strategy allows the player to win regardless of what moves the opponent will make [13]. Yet this notion, useful to analyse games such as chess, is too strong to investigate realistic communication. In most situations, agents are not restricted enough in their responses in order to "force" them to lose. For example, if the opponent always responds with a challenge (like in an exchange of speech acts: *claim* φ; *why* φ; *claim* ψ; *why* ψ; and so on), then there is no winning strategy for the proponent. Therefore, the first contribution of the paper is to introduce two weaker notions of dialogical strategies: a strategy giving a *chance* for success and a strategy giving a *particular degree* of chances for victory.

Strategies in games can be formally analysed using the Alternating-time Temporal Logic, ATL [2,8,7]. ATL specifies properties of game structures and uses alternating-time formulas to construct model-checkers in order to address problems such as receptiveness, realizability, and controllability. The logic introduces a path quantifier $\langle\langle A \rangle\rangle$ parameterized with the set A of players which ranges over all computations that the players in A can force the game into, irrespective of how the other players proceed. For example, the formula $\langle\langle A \rangle\rangle G\alpha$ expresses that the group of agents A has a strategy which ensure that always in the future, denoted by G, α holds. Yet the Alternating-time Temporal Logic is defined for any game and does not take into account the specifics of dialogue games. Moreover, it focuses on the notion of the winning strategy.

The solution to this problem is to apply a formalism which is specifically designed to analyse and verify dialogue games. To the best of the authors' knowledge, the only proposal of such a logic is the modal logic of Actions and Graded Beliefs, $\mathcal{AG}_n$ [4,5]. The aim of $\mathcal{AG}_n$ is to develop logic-based verification techniques for multi-agent systems in which agents communicate using dialogue-game based protocols. Originally, $\mathcal{AG}_n$ was introduced to express beliefs and persuasive actions of agents. Then, the logic was implemented to allow the analyst to investigate different properties of a multi-agent system using the methods of semantic and parametric verification. That is, the model checker allows the analyst to determine whether a given $\mathcal{AG}_n$ formula, which describes a property, is true at a given state of a given model, and to look for the valuations of an expression with some unknowns such that the obtained $\mathcal{AG}_n$ formula is true at a given state of a given model (see [6] for more details about a software tool Perseus). The second contribution of the paper is to extend the $\mathcal{AG}_n$ logic designed for reasoning about dialogue games with the elements of the ATL logic designed for reasoning about strategies in games and, as a result, to propose the ATL-like formalization of dialogical strategies (specifically, six modal operators for strategies in a dialogue games are introduced). In the future, it will be possible to implement this formalism in the Perseus model checker allowing the semantic and parametric verification of agents' strategies in a dialogue. As far as we are aware, there are no other logics for reasoning about dialogical strategies.

The paper is structured as follows. Section 2 describes a formal dialogue system that is used in the paper to model agent communication. Section 3 presents the main features of the $\mathcal{AG}_n$ logic. In Section 4, we discuss the types of strategies that might be adopted by agents during communication. Finally, Section 5 introduces formalization of the dialogical strategies using the frameworks of $\mathcal{AG}_n$ and ATL.

2 A Dialogue System for Argumentation

In this section we describe an example of a system in which persuasion dialogues can be formalized. This system was proposed by H. Prakken in [12]. We use it as a starting point for our further studies. Below, the main definitions are quoted.

A *dialogue system* for argumentation (dialogue system for short) is a pair $(\mathcal{L}; \mathcal{D})$, where $\mathcal{L}$ is a logic for defeasible argumentation and $\mathcal{D}$ is a dialogue system proper. A *logic for defeasible argumentation* $\mathcal{L}$ is a tuple $(L_t, R, Args, \rightarrow)$, where L_t (the topic language) is a logical language, R is a set of inference rules over L_t, $Args$ (the arguments) is a set of AND-trees of which the nodes are in L_t and the AND-links are inferences instantiating rules in R, and $\rightarrow$ is a binary relation of defeat defined on $Args$. For any argument A, $prem(A)$ is the set of leaves of A (its premises) and $conc(A)$ is the root of A (its conclusion).

A *dialogue system proper* is a triple $\mathcal{D} = (L_c; P; C)$ where L_c (the communication language) is a set of locutions, P is a protocol for L_c, and C is a set of effect (commitment) rules of locutions in L_c, specifying the effects of the locutions on the participants' commitments. The protocol for L_c is defined in terms of the notion of a dialogue, which in turn is defined with the notion of a move. The set M of *moves* is defined as $\mathbb{N} \times \{prop, opp\} \times L_c \times \mathbb{N}$, where the four elements of a move m are denoted by, respectively: $id(m)$ - the identifier of the move, $pl(m)$ - the player of the move, $s(m)$ - the speech act performed in m, $t(m)$ - the target of m. In [12], the following dialogue moves are considered: $claim(\alpha)$ – the speaker asserts that α is the case, $why(\alpha)$ – the speaker challenges α and asks for reasons why it would be the case, $concede(\alpha)$ – the speaker admits that α is the case, $retract(\alpha)$ – the speaker declares that he is not committed (any more) to α, $argue(A)$ – the speaker provides an argument A.

The set of *dialogues*, $M^{\leq \infty}$, is the set of all sequences $m_1, \ldots, m_i, \ldots$ from M such that each ith element in the sequence has identifier i, $t(m_1) = 0$, for all $i > 1$ it holds that $t(m_i) = j$ for some m_j preceding m_i in the sequence. The set of finite dialogues, denoted by $M^{<\infty}$, is the set of all finite sequences that satisfy these conditions. Note that the definition of dialogue implies that two speakers cannot speak at the same time. A *turntaking function* T is a function $T : M^{<\infty} \rightarrow 2^{\{prop,opp\}}$ such that $T(\emptyset) = \{prop\}$. A turn of a dialogue is a maximal sequence of stages in the dialogue where the same player moves.

When $T(d)$ is a singleton, the brackets will be omitted. Note that the definition alllows that more than one speaker has the right to speak next.

Rules of dialogical games are determined by the crucial element of the dialogue system, i.e., the protocol. Formally, a *protocol* on the set of moves M is a set $P \subseteq M^{<\infty}$ satisfying the condition that whenever d is in P, so are all initial sequences that d starts with. A partial function $Pr : M^{<\infty} \rightarrow 2^M$ is derived from P as follows: $Pr(d) =$ undefined whenever $d \notin P$; $Pr(d) = \{m : (d; m) \in P\}$ otherwise. The elements of $dom(Pr)$ (the domain of Pr) are called the legal finite dialogues. The elements of $Pr(d)$ are called the moves allowed after d. Every utterance from L_c can influence participants commitments. Results of utterances are determined by commitment rules which are specified as a function: $C : M^{<\infty} \times Agt \rightarrow 2^{L_t}$ for a participant $i \in Agt$ and a stage of a dialogue $d \in M^{<\infty}$. $C(d, i)$ denotes a player i's commitments at a stage of a dialogue d. If d is a legal dialogue and $Pr(d) = \emptyset$, then d is said to be a terminated dialogue.[1]

3   A Logic for the Dialogue System

For formalization of persuasion dialogues we propose to apply a logical system called Logic of Actions and Graded Beliefs, $\mathcal{AG}_n$ in short, which was introduced by Budzynska and Kacprzak in [4]. Its revised version for representation of Prakken's system was presented in [5]. This section presents this approach.

3.1   Beliefs of Agents

Beliefs of agents are modelled by means of doxastic relations. For every agent and every state of a model a doxastic relation determines states which the agent considers as possible current state of a system. Knowing how many states the agent considers and in how many of them some formula α is true, the degree of belief about α can be computed. For example, if an agent i has k_2 doxastic alternatives and in k_1 of them α holds, then we say that i believes α with degree $\frac{k_1}{k_2}$. It is denoted by $M!_i^{k_1,k_2}\alpha$. This formula is derived from a modality $M_i^k\alpha$ which intuitively says that α is satisfied in more than k doxastic alternatives of i. The degree an agent's belief may be changed by an action.

3.2   Commitments

The key concept of a dialogue system is the notion of commitment. In order to add commitments to $\mathcal{AG}_n$ logic, we define a commitment function C which for every state of a model and every agent assigns a set of formulas. Moreover we assume that L_t and L_c are represented by means of the $\mathcal{AG}_n$ logic. Furthermore, following [15] we assume that the union of the commitment stores can be viewed as some state of the dialogue. Thus, in our specification commitments are assigned to a state rather than a sequence of dialogue moves (as assumed in [12]).

---

[1] For present purposes a more detailed definitions are not needed. For the full details the reader is referred to [12].

The commitment operator is denoted by $\mathbf{C}$. A formula $\mathbf{C}_i\alpha$ intuitively says that α is a commitment of agent i.

3.3   Persuasion Actions

In our approach, two types of actions are taken into account: physical and verbal. Physical actions modify states of a model and as a result affect agent's beliefs. Modeling of verbal actions is a bit complex. The goal of a verbal action is not to change a state of a model but only doxastic alternatives or commitments of agents assigned to this state. Thus the whole model is transformed to a new one. In the updated model, states stay the same but doxastic relations or commitment functions are modified. In this way, physical actions influence agents' environment while verbal actions influence agents' perception of this environment. The verbal actions we use are: $claim(\alpha)$, $why(\alpha)$, $concede(\alpha)$, $retract(\alpha)$, $argue(A)$, where α is a formula and $A \in Args$ is an argument. The possible result of the execution of a sequence of actions P is expressed by the formula $\Diamond(i : P)\alpha$, where α is a formula which describes some property.

3.4   Formal Syntax and Semantics

Let $Agt = \{1, \ldots, n\}$ be a set of names of *agents*, V_0 be a set of *propositional variables*, Π_0^{ph} a set of *physical actions*, and Π_0^v a set of *verbal actions*. Further, let ; denote a programme connective which is a sequential composition operator. It enables to compose *schemes of programs* defined as finite sequences of atomic actions: $a_1; \ldots; a_k$. Intuitively, the program $a_1; a_2$ for $a_1, a_2 \in \Pi_0^{ph}$ means "Do a_1, then do a_2". The set of all schemes of physical programs we denote by Π^{ph}. In similar way, we define a set Π^v of schemes of programs constructed over Π_0^v.

The set of F all *well-formed expressions* of the extended $\mathcal{AG}_n$ is given by the following Backus-Naur form:

$$\alpha ::= p|\neg\alpha|\alpha \vee \alpha|M_i^k\alpha|\Diamond(i : P)\alpha|\mathbf{C}_i\alpha,$$

where $p \in V_0$, $k \in \mathbb{N}$, $i \in Agt$, $P \in \Pi^{ph}$ or $P \in \Pi^v$. We also use abbreviations: $\Box(i : P)\alpha$ for $\neg\Diamond(i : P)\neg\alpha$, $B_i^k\alpha$ for $\neg M_i^k\neg\alpha$, $M!_i^k\alpha$ where $M!_i^0\alpha \Leftrightarrow \neg M_i^0\alpha$, $M!_i^k\alpha \Leftrightarrow M_i^{k-1}\alpha \wedge \neg M_i^k\alpha$, if $k > 0$, and $M!_i^{k_1,k_2}\alpha$ for $M!_i^{k_1}\alpha \wedge M!_i^{k_2}(\alpha \vee \neg\alpha)$. By a *semantic model* we mean a Kripke structure $M = (S, RB, I^{ph}, v, C)$ where

- S is a non-empty set of states (the universe of the structure),
- RB is a doxastic function which assigns to every agent a binary relation, $RB : Agt \longrightarrow 2^{S \times S}$,
- I^{ph} is an interpretation of physical actions, $I^{ph} : \Pi_0^{ph} \longrightarrow (Agt \longrightarrow 2^{S \times S})$,
- $C : S \times Agt \longrightarrow 2^F$ is a commitment function,
- v is a valuation function, $v : S \longrightarrow \{\mathbf{0}, \mathbf{1}\}^{V_0}$.

Function I^{ph} can be extended in a simple way to define interpretation of any program scheme. Let $I_{\Pi^{ph}}^{ph} : \Pi^{ph} \longrightarrow (Agt \longrightarrow 2^{S \times S})$ be a function such that

$I^{ph}_{\Pi^{ph}}(P_1; P_2)(i) = I^{ph}_{\Pi^{ph}}(P_1)(i) \circ I^{ph}_{\Pi^{ph}}(P_2)(i) = \{(s, s') \in S \times S : \exists_{s'' \in S} ((s, s'') \in I^{ph}_{\Pi^{ph}}(P_1)(i)$ and $(s'', s') \in I^{ph}_{\Pi^{ph}}(P_2)(i))\}$ for $P_1, P_2 \in \Pi^{ph}$ and $i \in Agt$.

Furthermore, let $\mathcal{CM}$ be a class of models and $\mathcal{CMS}$ be a set of pairs (M, s) where M $\in \mathcal{CM}$ and s is a state of the model M. On the set $\mathcal{CMS}$ we define a function I^v which is an interpretation for verbal actions: $I^v : \Pi^v_0 \longrightarrow (Agt \longrightarrow 2^{\mathcal{CMS} \times \mathcal{CMS}})$. Interpretation $I^v_{\Pi^v}$ of all verbal programs is defined similarly to the function $I^{ph}_{\Pi^{ph}}$.

The *semantics* of formulas is defined with respect to $\mathcal{M}$, i.e., for a given M $= (S, RB, I^{ph}, v, C)$ and $s \in S$ the Boolean value of the formula α is denoted by M, $s \models \alpha$ and is defined inductively as follows:

M, $s \models p$ iff $v(s)(p) = \mathbf{1}$, for $p \in V_0$,

M, $s \models \neg\alpha$ iff M, $s \not\models \alpha$,

M, $s \models \alpha \vee \beta$ iff M, $s \models \alpha$ or M, $s \models \beta$,

M, $s \models M^k_i\alpha$ iff $|\{s' \in S : (s, s') \in RB(i)$ and M, $s' \models \alpha\}| > k$, $k \in \mathbb{N}$,

M, $s \models \Diamond(i : P)\alpha$ iff $\exists_{s' \in S} ((s, s') \in I^{ph}_{\Pi^{ph}}(P)(i)$ and M, $s' \models \alpha)$ for $P \in \Pi^{ph}$ or $\exists_{(M',s') \in \mathcal{CMS}} (((M, s), (M', s')) \in I^v_{\Pi^v}(P)(i)$ and M', $s' \models \alpha)$ for $P \in \Pi^v$,

M, $s \models C_i\alpha$ iff $\alpha \in C(s, i)$.

3.5   Representation of Speech Acts

Based on commitment rules (cf. [12]) we introduce a specification of dialogue actions. Since speech acts are verbal actions they are intended to influence the beliefs or commitments of agents. At the same time, states of a model remain unchanged. Therefore a dialogue action, like any verbal action in our logic, moves a system from a model M to a new model M'. For instance, an action $claim(\alpha)$ performed at a state s of a model M moves a multi-agent system to a model M' in which a new commitment function C' is defined in such a way that the new set of commitments of the performer of the action at s equals to the old one enriched with α. Because in [12] the change of commitments do not need to be a subset of beliefs, in the new model M' the doxastic relation may stay unchanged. Below we give a formal definition of function I^v, i.e., interpretation of dialogue actions. Let M $= (S, RB, I^{ph}, v, C)$. Let us start with an interpretation of *claim*:

$((M, s), (M', s)) \in I^v(claim(\alpha))(i)$ iff   M' $= (S, RB, I^{ph}, v, C')$ where $C'(s, i) = C(s, i) \cup \{\alpha\}$ and $C'(s', i') = C(s', i')$ for $s' \neq s$ or $i' \neq i$.

At state s, agent i adds the formula α to the set of its commitments and all other commitments stay unchanged. The interpretation of *concede* is exactly the same, i.e.:

$((M, s), (M', s)) \in I^v(concede(\alpha))(i)$ iff   M' $= (S, RB, I^{ph}, v, C')$ where $C'(s, i) = C(s, i) \cup \{\alpha\}$ and $C'(s', i') = C(s', i')$ for $s' \neq s$ or $i' \neq i$.

The interpretation of *retract* is as follows:

$((M, s), (M', s)) \in I^v(retract(\alpha))(i)$ iff   M' $= (S, RB, I^{ph}, v, C')$ where $C'(s, i) = C(s, i) \backslash \{\alpha\}$ and $C'(s', i') = C(s', i')$ for $s' \neq s$ or $i' \neq i$.

At state s, agent i deletes the formula α from the set of its commitments and all other commitments stay unchanged. The interpretation of *argue* is as follows:

$$((\mathrm{M}, s), (\mathrm{M}', s)) \in I^v(argue(A))(i) \quad \text{iff} \quad \mathrm{M}' = (S, RB, I^{ph}, v, C') \text{ where}$$
$$C'(s, i) = C(s, i) \cup prem(A) \cup conc(A) \quad \text{and}$$
$$C'(s', i') = C(s', i') \text{ for } s' \neq s \text{ or } i' \neq i.$$

At state s, agent i adds the premises and the conclusion of an argument A to the set of its commitments and all other commitments stay unchanged. The interpretation of *why* is as follows:

$$((\mathrm{M}, s), (\mathrm{M}', s)) \in I^v(why(\alpha))(i) \quad \text{iff} \quad \mathrm{M}' = \mathrm{M}.$$

In this case, commitments are not changed.

4   Types of Strategies in Persuasion Dialogues

The aim of our research is to formally analyze dialogical systems in order to learn about strategies of agents in dialogues. More specifically, we are interested in conditions which can cause victory or a failure of agents in a game. Therefore we need to verify systems with respect to questions about how successful different strategies are.

4.1   Sequence of Dialogue Moves Leading to Victory

The simple question regarding an agent's victory in a dialogue would be to ask if a given sequence of moves allows the agent to accomplish his goal. In order to verify such a question, we first need to specify the notion of a victory of a persuasion dialogue. Let $win(i)$ mean that i is a winner of a given dialogue game. As noted above, the notion of a victory can be understood in different manners depending on the applications we consider. One possible specification is to assume that a proponent i is the winner if the opponent has conceded i's main claim and an opponent is the winner if the proponent i has retracted i's main claim [12]. Then, if we consider a dialogue game with the topic t (i.e. a conflict formula), the proponent $i \in prop(t)$, the opponent $\bar{i} \in opp(t)$, played in accordance with the protocol P, then:

- $win(i)$ is true in a state, in which $\mathbf{C}_{\bar{i}}(t)$ holds, while
- $win(\bar{i})$ is true in a state, in which $\neg\mathbf{C}_i(t)$ holds.

Using an assumed specification for the notion of a victory, we can ask: is it possible that after performing a dialogue d between i and $\bar{i}$ played in accordance with the protocol P, it will be the case that the proposition $win(i)$ will hold, or: what dialogue has to be performed such that it will be possible that $win(i)$ will hold. In the first case, an agent may ask if a specific sequence, e.g. *claim p; why p; p since q; ...; concede s*, can lead him to win a dialogue. In other words, he may ask if he can be a winner when a dialogue proceeds according to a specific scenario.

Perhaps the more interesting question would be to ask *what* sequence of moves allows an agent to win a dialogue game. This question requires our model checker to perform parametrical verification. It means that Perseus searches for a legal dialogue (i.e. a dialogue played according to a given protocol) such that it is possible that after performing it the proposition $win(i)$ (or $win(\bar{i})$) will hold. Knowing that a given dialogue sequence is "victorious" allows an agent to plan how he should play the dialogue game in order to win it.

Yet this type of question has a strong limitation. Unlike other types of sequence of actions, a dialogue always consists of actions executed not only by one agent, but also by his adversary. Intuitively it means that part of a sequence is not under control of a given agent. As a result, even though i knows that a particular sequence leads him to the victory, this sequence may not be performed in a dialogue, since $\bar{i}$ may execute the action allowed by a dialogue protocol, but other than considered by i. Say that a sequence *claim p*; *why p*; *p since q*; ...; *concede s* allows i to win a dialogue. However, in the second move $\bar{i}$ may execute *claim ¬p* instead of *why p*. Consequently, i's knowledge that the sequence is victorious becomes useless in this case. Therefore, we need a stronger notion that will allow an agent to reason about the victory regardless of what action his opponent will execute. In the next section, we consider the notions of a strategy and a winning strategy.

4.2   Winning Strategy

As mentioned above, a strategy for an agent i can be defined as a function from the set of all finite legal dialogues in which i is to move into L_c [13]. Intuitively, i has a strategy if he has a plan of how to react to any move of his adversary. Say that the first move *claim p* is performed by i. At this state, i considers how he will response after all possible moves that $\bar{i}$ is allowed to make at the next stages of a dialogue. In particular, he may plan that at the subsequent stage (i.e. after *claim p*) if $\bar{i}$ executes *why p*, then his response will be: *p since q* (instead of, e.g., *retract p*), if $\bar{i}$ executes *claim ¬p*, then his response will be: *why ¬p* (instead of, e.g., *concede ¬p*), and so on.

An agent may want to know if a strategy that he adopted guarantees him the victory in a given dialogue game regardless of what actions his opponent will perform. This type of strategy is called a winning strategy. A strategy is a winning strategy for i if in every dialogue played according to this strategy i accomplishes his dialogue goal [13]. Knowing that a given strategy is a winning one allows an agent to choose and adopt it and, in consequence, to win.

The question about a winning strategy has some limitations when applied to the systems of persuasion dialogue. It is well-known (see e.g. [12, p. 1021]) that playing according to a protocol a player may avoid losing simply by never giving in, e.g. an opponent may repeat *why α* as a response to any assertion that the proponent performs, such as *claim α* or *β since α*. Figure 1 shows the fragment of a dialectical tree constructed according to the protocol from [12]. In such a case, there is no winning strategy for the proponent, since no matter how good his responses will be, there will be a path where he cannot win, as long as his opponent will infinitely keep repeating *why* α (e.g. *claim p*; *why p*; *p since q*; *why q*; *q since r*; *why r*, and so on).

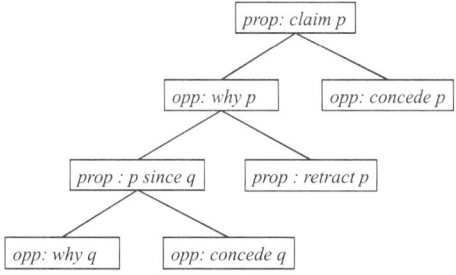

Fig. 1. The fragment of a dialectical tree

In such cases, the notion of winning strategy is too rigid in this sense that an agent cannot choose a strategy that has this feature (since there does not exist any). In other words, if a protocol allows for such a game in which the response follows the pattern "why why why", there is no possibility to construct a winning strategy. Thus, in those cases we need a slightly "weaker" type of strategy, which will allow an agent to reason not about the guarantee but about the possibility of victory. In the next section, we consider the notion of a strategy, which gives an agent chance for success.

4.3   Strategy Giving Chance for Success

An agent may want to know if a strategy that he adopted gives him a chance to be successful in a given dialogue game. Intuitively, a strategy gives i chance for success if there is a dialogue game played according to this strategy such that i accomplishes his dialogue goal. Knowing that the strategy has this feature allows an agent to make decision about which strategy he should adopt in order to have a chance to win. Even though the agent is not sure that he will win, the information that one strategy can bring him success and the other cannot is better than no information.

However, the question about a strategy giving chance for success has some limitations. Say that an agent knows that ten strategies allows him to be victorious. How can he decide which strategy to choose? In the next section, we consider the notion of a strategy, which gives an agent chance for success in a certain degree.

4.4   Degree of Chance of Success

An agent may wish to know in how many cases a strategy gives him a chance to be victorious. Assume a class of dialogue games in which there is a finite number of possible game's scenarios. Let k_2 be the number of all dialogues played according to a given strategy, and k_1 be a number of dialogues played according to this strategy in which a given agent i accomplishes his dialogue goal. If k_1 and k_2 are finite, then we say that this strategy gives i chance for success in a degree $\frac{k_1}{k_2}$. If they are infinite, the degree of chance for success is not defined. Knowing that ten strategies allows him to be victorious and knowing their degrees of chance for success allows an agent to choose among them and, in consequence, to maximize his chance to win.

4.5   Types of Questions about Victory

In summary, we are interested in verification related to six types of questions:

- Is there a strategy that allows an agent to win a dialogue game?
- Is there a strategy that allows an agent to win with a certain frequency (with a certain degree of chance for success)?
- Is there a winning strategy?
- Does a given strategy allow an agent to win a dialogue game?
- How high the chance for success is?
- Is a given strategy winning?

5   Formalization of Dialogical Strategies

In this section we introduce new strategy operators. Their interpretation is based on a formal notion of a computation and a strategy. Formulas with these operators describe properties of dialogical systems which express that a specific strategy may lead to a victory, or leads to a victory with some degree, or always leads to a victory, i.e. is a winning strategy, etc. The operators are inspired by operators of ATL logic.

5.1   Basic Definitions

Let

$$\delta : \mathcal{CM} \times S \times Agt \rightarrow 2^{(\Pi^{Ph} \cup \Pi^v)}$$

be a function mapping a triple consisting of a model, a state of this model and an agent to a set of actions. These actions are assumed to be actions which the agent can perform next. In fact this function determines *transition function*, i.e. indicates models and states of these models reachable from a given state of a given model by a given agent.

Definition 1. *A **computation** is a sequence*

$$(M_0, s_0), (M_1, s_1), (M_2, s_2), \ldots$$

such that for every $k \geq 0$, there exists an action a_k and an agent i_k such that $a_k \in \delta(M_k, s_k, i_k)$ and $((M_k, s_k), (M_{k+1}, s_{k+1})) \in I(a_k, i_k)$ where I is the interpretation of action a_k, i.e., $I = I^{ph}$ if a_k is a physical action and $I = I^v$ if a_k is a verbal action.

Intuitively by a computation we mean a sequence of pairs (M_k, s_k), a model and a state of this model, such that for every position k, $(\mathrm{M}_{k+1}, s_{k+1})$ is a result of performing an action a_k by an agent i_k at the state s_k of the model M_k.

Definition 2. *By a **strategy** for an agent i we call a mapping $f_i : M^{<\infty} \to 2^M$ which assigns to every finite dialogue $d = m_0, m_1, \ldots, m_k \in M^{<\infty}$ in which it is i's turn, i.e., $i \in T(d)$, a move $m \in M$ such that $m \in Pr(d)$.*

In other words, a strategy function returns a move which is allowed by the protocol P after a dialogue d where i is to move. We say that a dialogue $d = m_0, m_1, \ldots$ is consistent with a strategy f_i iff for every $k \geq 1$ if $i = pl(m_k)$ then $m_k \in f_i(m_0, \ldots, m_{k-1})$ and for $k = 0$ if $i = pl(m_k)$ then $m_k \in f_i(\emptyset)$, i.e., every move of agent i is determined by the function f_i.

Next, we define the **outcomes** of f_i, i.e., a set of computations which are consistent with this strategy. Let $\lambda = (\mathrm{M}_0, s_0), (\mathrm{M}_1, s_1), (\mathrm{M}_2, s_2), \ldots$ be a computation, then

$$\lambda \in out((\mathrm{M}, s), f_i) \quad iff \quad (\mathrm{M}_0, s_0) = (\mathrm{M}, s) \quad and$$

there exists a dialogue $d = m_0, m_1, \ldots$ consistent with f_i such that

$$\text{for every } k \geq 0, \quad s(m_k) \in \delta(\mathrm{M}_k, s_k, pl(m_k))$$

and

$$((\mathrm{M}_k, s_k), (\mathrm{M}_{k+1}, s_{k+1})) \in I(pl(m_k), s(m_k)).$$

Intuitively, a computation is consistent with a strategy if it is determined by a dialogue consistent with the strategy.

5.2 Strategy Operators

In this section we introduce strategy operators. Let $d = m_0, m_1, m_2, \ldots$ be a dialogue and $\Diamond(d)\alpha$ will be a short for

$$\Diamond(pl(m_0) : s(m_0))\Diamond(pl(m_1) : s(m_1)) \ldots \Diamond(pl(m_k) : s(m_k))\alpha.$$

Then the formula

$$\Diamond(d)\alpha$$

expresses that it is possible that after the performing of the dialogue d, α holds. If α says t then the formula expresses that d may terminate with the success. This property is important, but we need to know more about a dialogue system.

It is more informative for an agent to know whether a proponent has a strategy which leads him to a successful state. In other words, we ask whether the proponent knows such responses for his adversary which *may* result in the proponent being able to achieve his goal. To express such a property we need to introduce a new operator:

$$\langle i \rangle G\alpha.$$

It says that there *exists* a strategy of i and there *exists* a computation consistent with this strategy such that in all states of this computation α is true. Formally:

$M, s \models \langle i \rangle G\alpha$ iff there exists a strategy f_i such that for some computation $\lambda = (M_0, s_0), (M_0, s_0), \cdots \in out((M, s), f_i)$, and for all $k \geq 0$, we have $(M_k, s_k) \models \alpha$.

Given a finite set of computations consistent with some strategy we can compute to what degree this strategy is successful. It will be measured by a fraction $\frac{k_1}{k_2}$ where k_2 is the number of all computations and k_1 is the number of computations in which every state satisfy α. Then the formula

$$\langle i \rangle^k G\alpha$$

expresses that agent i has such a strategy for which $\frac{k_1}{k_2} \geq k$. The value $\frac{k_1}{k_2}$ will be called a degree of a strategy.

The most useful are such strategies as will *always* lead to success. This property is expressed by means of the following operator

$$\langle\!\langle i \rangle\!\rangle G\alpha.$$

It expresses that there *exists* such a strategy which *always* leads to success regardless of the opponent actions, i.e. the proponent has a *winning strategy*:

$M, s \models \langle\!\langle i \rangle\!\rangle G\alpha$ iff there exists a strategy f_i such that for all computations $\lambda = (M_0, s_0), (M_1, s_1), \cdots \in out((M, s), f_i)$, and for all positions $k \geq 0$, we have $M_k, s_k \models \alpha$.

Let f be a strategy of agent i. Sometimes we need to check what features the strategy has. So we introduce the operator

$$\langle i \rangle_f G\alpha.$$

Intuitively it says that for the strategy f there *exists* a computation consistent with this strategy such that in all states of this computation α is true. Formally:

$M, s \models \langle i \rangle_f G\alpha$ iff for some computation $\lambda = (M_0, s_0), (M_1, s_1), \cdots \in out((M, s), f)$, and for all positions $k \geq 0$, we have $M_k, s_k \models \alpha$.

Next the operator

$$\langle i \rangle_f^k G\alpha$$

expresses that the strategy f of agent i has degree $\frac{k_1}{k_2} \geq k$.

The last operator we use is

$$\langle\!\langle i \rangle\!\rangle_f G\alpha.$$

It expresses that the strategy f is a winning strategy. Formally:

$M, s \models \langle\!\langle i \rangle\!\rangle_f G\alpha$ iff for all computations $\lambda = (M_0, s_0), (M_1, s_1), \cdots \in out((M, s), f)$, and for all positions $k \geq 0$, we have $M_k, s_k \models \alpha$.

Similar to ATL logic we can consider also strategy formulas $\langle\!\langle i \rangle\!\rangle X\alpha$ and $\langle\!\langle i \rangle\!\rangle \alpha U \beta$ with the following semantics:

$M, s \models \langle\!\langle i \rangle\!\rangle X\alpha$ iff there exists a strategy f_i such that for all computations $\lambda = (M_0, s_0), (M_1, s_1), \cdots \in out((M, s), f_i)$, we have $M_1, s_1 \models \alpha$,

$M, s \models \langle\!\langle i \rangle\!\rangle \alpha U \beta$ iff there exists a strategy f_i such that for all computations $\lambda = (M_0, s_0), (M_1, s_1), \cdots \in out(s, f_i)$, there exists a position $k \geq 0$ such that $M_k, s_k \models \beta$ and for all positions $0 \leq j < k$, we have $M_j, s_j \models \alpha$.

The formulas $\langle i \rangle X\alpha$, $\langle i \rangle^k X\alpha$, $\langle i \rangle_f X\alpha$, $\langle i \rangle_f^k X\alpha$, $\langle\!\langle i \rangle\!\rangle_f X\alpha$, $\langle i \rangle \alpha U \beta$, $\langle i \rangle^k \alpha U \beta$, $\langle i \rangle_f \alpha U \beta$, $\langle i \rangle_f^k \alpha U \beta$, $\langle\!\langle i \rangle\!\rangle_f \alpha U \beta$ can be defined analogously.

5.3   Examples of Properties

The most important property, which we would like to verify in a given system, is whether there exists a winning strategy in a game for a given agent i, i.e., whether there exists a strategy which enures that i will win for sure. This property is expressed by the formula:

$$\langle\!\langle i \rangle\!\rangle \ true \ U \ win(i).$$

If a strategy is not a winning strategy but leads to a success with degree higher than 0.75 then it can be expressed by the following formula:

$$\langle i \rangle^{0.75} \ true \ U \ win(i).$$

The success of the agent i can mean various events. For example, the agent i will win the game if its adversary believes a sentence t with degree $\frac{3}{4}$. The following formula expresses that agent i has a chance to achieve this success:

$$\langle\!\langle i \rangle\!\rangle \ true \ U \ (M!_{\overline{i}}^{3,4})t.$$

Similarly, the agent i is a winner if i's adversary is committed to a sentence t. The property which says that there is a strategy for i which ensures this kind of victory is expressed by the formula:

$$\langle\!\langle i \rangle\!\rangle \ true U \ C_{\overline{i}}(t).$$

6   Conclusions

The paper addresses the problem of analysis of persuasion dialogue games. It proposes a formal framework which allows for specification of properties of dialogical systems concerning winning strategies and abilities. The presented approach applies verification techniques on modal logics, specifically variations of the Alternating-time temporal logic ATL, to games for argumentation. A language with new strategy operators is introduced and studied. It extends the language of $\mathcal{AG}_n$ logic which is a basis for the software tool Perseus designed for automated verification of properties of persuasion processes conducted in multi-agent systems.

Acknowledgments. We gratefully acknowledge the support of the Polish National Science Center under grant 2011/03/B/HS1/04559.

References

1. Foundation for Intelligent Physical Agents (FIPA). FIPA communicative act library specification (2002), http://www.fipa.org
2. Alur, R., Henzinger, T.A., Kupferman, O.: Alternating-time temporal logic. Journal of the ACM 49(5), 672–713 (2002)
3. Black, E., Hunter, A.: An inquiry dialogue system. Autonomous Agents and Multi-Agent Systems 19(2), 173–209 (2009)
4. Budzyńska, K., Kacprzak, M.: A logic for reasoning about persuasion. Fundamenta Informaticae 85, 51–65 (2008)
5. Budzynska, K., Kacprzak, M.: Formal framework for analysis of agent persuasion dialogue games. In: Proc. of CS&P, pp. 85–96 (2010)
6. Budzyńska, K., Kacprzak, M., Rembelski, P.: Perseus. software for analyzing persuasion process. Fundamenta Informaticae 93(1-3), 65–79 (2009)
7. Bulling, N., Dix, J., Jamroga, W.: Model checking logics of strategic ability: Complexity. In: Specification and Verification of Multi-Agent Systems, pp. 125–159 (2010)
8. Jamroga, W., van der Hoek, W.: Agents that know how to play. Fundamenta Informaticae 63(2-3), 185–219 (2004)
9. Larson, K., Rahwan, I.: Welfare properties of argumentation-based semantics. In: Proceedings of the 2nd COMSOC (2008)
10. Maudet, N., Chaib-draa, B.: Commitment-based and dialogue-game based protocols: new trends in agent communication languages. The Knowledge Engineering Review 17(2), 157–179 (2002)
11. McBurney, P., Parsons, S.: Dialogue games in multi-agent systems. Informal Logic 22(3), 257–274 (2002)
12. Prakken, H.: Coherence and flexibility in dialogue games for argumentation. Journal of Logic and Computation (15), 1009–1040 (2005)
13. Prakken, H.: Formal systems for persuasion dialogue. The Knowledge Engineering Review 21, 163–188 (2006)
14. Rahwan, I., Larson, K., Tohme, F.: A characterisation of strategy-proofness for grounded argumentation semantics. In: Proceedings of the 21st IJCAI (2009)
15. Sklar, E., Parsons, S.: Towards the application of argumentation-based dialogues for education. In: Proceedings of AAMAS 2004, vol. 3, pp.1420–1421 (2004)
16. Walton, D.N., Krabbe, E.C.W.: Commitment in Dialogue: Basic Concepts of Interpersonal Reasoning. State University of N.Y. Press (1995)

Low–Cost Computer Vision Based Automatic Scoring of Shooting Targets

Jacek Rudzinski and Marcin Luckner

Warsaw University of Technology, Faculty of Mathematics and Information Science,
pl. Politechniki 1, 00–661 Warsaw, Poland
rudzinskij@student.mini.pw.edu.pl,
mluckner@mini.pw.edu.pl
http://www.mini.pw.edu.pl/~lucknerm/en/

Abstract. This paper introduces an automatic scoring algorithm on shooting target based on computer vision techniques. As opposed to professional solutions, proposed system requires no additional equipment and relies solely on existing straightforward image processing such as the Prewitt edge detection and the Hough transformation. Experimental results show that the method can obtain high quality scoring. The proposed algorithm detects holes with 99 percent, resulting in 92 percent after eliminating false positives. The average error on the automatic score estimation is 0.05 points. The estimation error for over 91 percent holes is lower than a tournament–scoring threshold. Therefore the system can be suitable for amateur shooters interested in professional (tournament-grade) accuracy.

Keywords: Computer vision, Hough transform, Pattern recognition, Score estimation.

1   Introduction

This paper describes a low–cost computer vision based system for shooting target scoring. Typically, automatic scoring of shooting target is done by expensive equipment such as shooting targets scanners [6], optical evaluation systems [2], electronic shooting targets [13], or acoustic systems [3].

A cheaper alternative can be created using computer vision based automatic scoring systems. However, existing systems also require special equipment such as high–resolution digital camera [1], DH–CG320 capture card [8], or a laser [12]. Moreover, these solutions are designed as stationary systems for professional shooters.

Our aim was a creation of a professional system that could be used by amateur shooters. Such system should estimate the score on a level accepted by an International Shooting Sport Federation (ISSF) as presented in Fig. 1.

Since the solution is proposed to amateur shooters, the hardware requirements should be fulfilled by any consumer grade cameras or mobile devices capable of capturing image with at least 0.5 Mpix resolution.

M. Graña et al. (Eds.): KES 2012, LNAI 7828, pp. 185–195, 2013.

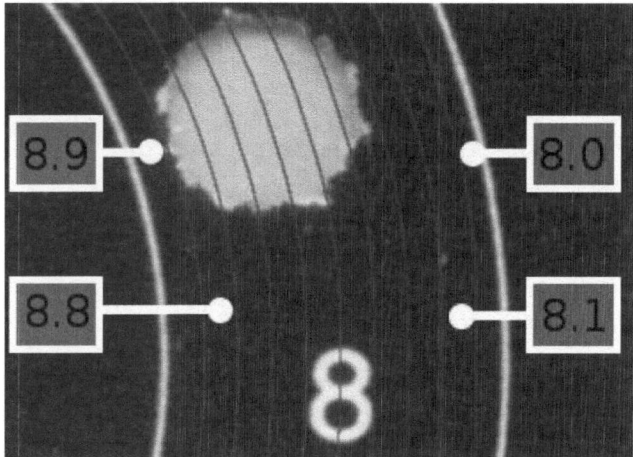

Fig. 1. A professional scoring with a tournament precision. The score is 8.8 in this case.

The described system consists of three steps. The first one is a target detection, which is described in section 2. The next step is the detection of holes, presented in section 3. The last step, which is assigning score values to detected holes is demonstrated in section 4.

The algorithm was tested on several amateur photographs of shooting targets and section 5 contains the complete experimental results. The paper is concluded in section 6.

2   Target Detection

The first step of target detection involves finding of the depicted target's components. The main element of the target is a bull–eye, which is the central black circle on a white background. Surrounding the center, there is a number of black (closer) and white (surrounding) circles that define scoring sections.

The minimal conditions for successful detection are that the whole bull–eye is visible and the optical axis tilted by no more than 30 degrees from the normal of the target plane.

2.1   Bull–Eye Detection

Because of inconsistent illumination and other factors resulting from blurred binary projection, simple binarization filters are insufficient for detecting target's components. Instead an algorithm extracting a list of segments is proposed.

The algorithm runs as follows. Firstly, the image resolution is reduced. Next, the Prewitt operator is used to detect edges. Finally, cohesive areas are detected by a flood fill and saved for future analysis.

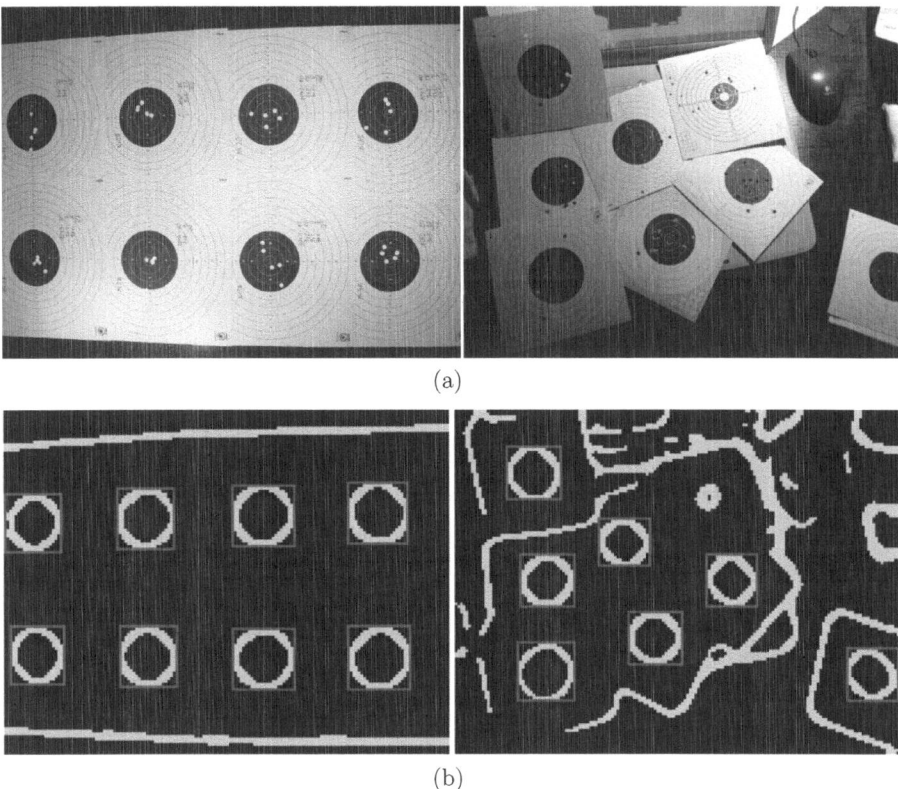

(a)

(b)

Fig. 2. (a) Amateur photos of targets (b) Results of bull–eyes detection algorithm. Detected bull-eyes are marked with squares.

Before proceeding with objects' analysis an additional feature, γ_p is calculated for each pixel p. It is defined as the brightness difference between the pixel and its eight–point neighborhood $P = \{p_1 \dots p_8\}$ as

$$\gamma_p = \max_{p_x \in P} |\text{lum}(p) - \text{lum}(p_x)|, \tag{1}$$

where lum returns a value (brightness) component of HSV model.

Now, each object from the list must satisfy the following two conditions to be recognized as a bull–eye. Each pixel lying on an edge must have it's brightness greater than the calculated maximum difference: (1):

$$\text{lum}(p) - \gamma_p > 0, \tag{2}$$

If almost all pixels satisfy the condition then the shape is a potential bull–eye. It will be accepted on the condition that its width and height are similar:

$$\frac{\max(W, H)}{\min(W, H)} < 2, \tag{3}$$

where W and H are width and height respectively. The results of the algorithm applied to amateur photos in Fig. 2(a) are shown in Fig. 2(b). Detected bull–eyes are marked with squares.

A detected bull–eye brings local information about a threshold for binarization. Points inside the bull–eye are dark whereas, points on the edge are light. For both sets of points average values of colors are calculated. The following procedure is run for each component of color model individually. Firstly, parameters of normal distribution are estimated for both sets. All points that lie farther than two standard deviations from the mean are removed to eliminate the influence of light elements on dark color estimation on a bull–eye. The threshold for binarization is set halfway between averages of dark and light points sets.

After the binarization, Thr set of points on the bull–eye's edge will be used as the source of information about geometry of rings projections.

2.2   Rings Detection

The set of points on the bull–eye edge determines the approximate shape of the main ring. After corrections, selected points can be used for the parametrization of the ring projection.

Correction. The bull–eye was detected in the low–resolution image, while the ring geometry should be detected in the original photo. For that reason the detected points are only approximate. It can be assumed that a point from a ring lies on the line determined by approximate point and the center of the target. Moreover, a distance between the points should be no more than $\frac{1}{6}$ of the distance between rings as it is shown in Fig. 3.

These conditions define a segment that includes the ring point. For the segment, the value V from HSV color model is calculated in each point. The segment has the minimum, the maximum, and the average values labeled as V_{min}, V_{max}, and V_{avg} respectively. For a given threshold $\alpha \in [0, 1]$ a point from the ring p_{ring} is calculated as

$$p_{ring} = \begin{cases} p_{min} & \text{if } \kappa > 1 - \alpha \\ p_{max} & \text{if } \kappa < \alpha \\ p_{diff} & \text{if } \alpha < \kappa > 1 - \alpha, \end{cases} \tag{4}$$

where

$$\kappa = \frac{V_{avg} - V_{min}}{V_{max} - V_{min}}. \tag{5}$$

The selected point can be the most brightnest p_{max}, the darkest p_{min}, or of the highest difference between neighbors p_{diff}.

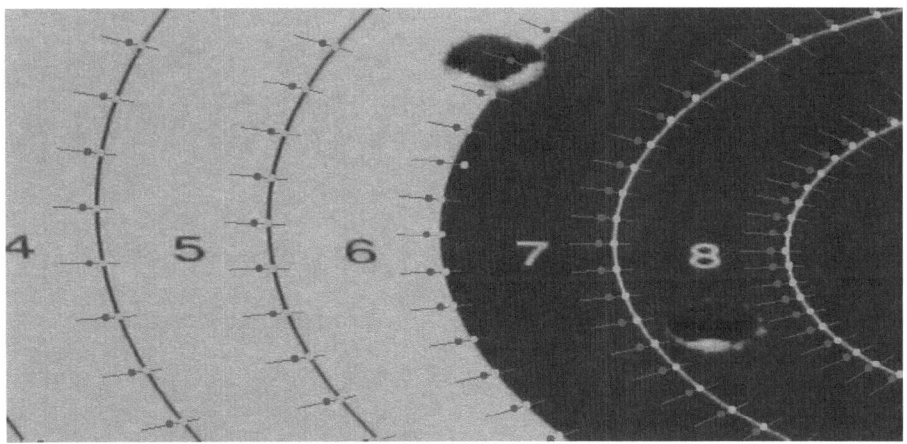

Fig. 3. Estimated points (dark) and actual rings' points (light)

Selection. Some points from the set can be taken as a part of the ring by a mistake. To avoid such situation, several ellipses are generated from random subsets of points [9] using the following algorithm:

An ellipse e based on m random points is created [10,11]. Next, for each ellipse the following steps are completed

1. All points in the nearest neighborhood of ellipse's center increase their counters. The nearest neighborhood is defined by d, distance in the Manhattan metric.
2. If a point with maximal counter is changed then go to the next step.
3. If distance between the last point with maximal counter and a new one is lower than d then add the ellipse e to a result list. In other case, remove all ellipses from the result list.

Newly created ellipses are evaluated with distance criteria. A subset with minimal variation of distances between points and foci should be selected as a base for the final ellipse.

Parametrization. An ellipse is given by the following equation:

$$F(x,y) = ax^2 + bxy + cy^2 + dx + ey + f = 0. \tag{6}$$

The parameters of the ellipse can be found using the least squares method [10]. However, this method is numerically unstable and a stable solution was used instead [11]. A MATLAB implemetation is given below.

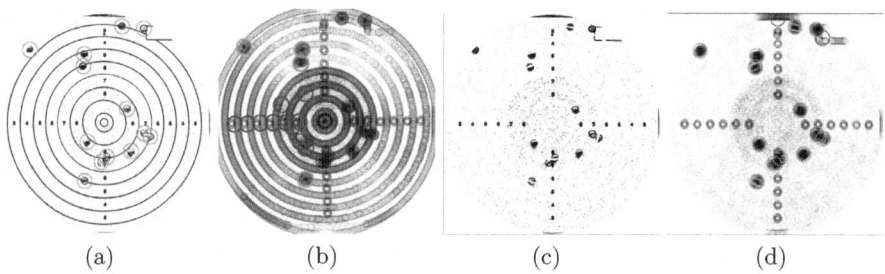

| (a) | (b) | (c) | (d) |

Fig. 4. Holes detection. 4(a) A result of the Prewitt operator. 4(b) Hough transformation results with a significant number of false positives. 4(c) A result of the Prewitt operator with erased rings. 4(d) Hough transformation results with a reduced number of false positives.

Ellipse parametrization

```
function a = fit_ellipse(x, y)
D1 = [x . 2, x .* y, y . 2]; % quadratic part of the
design matrix
D2 = [x, y, ones(size(x))]; % linear part of the design
matrix
S1 = D1 * D1; % quadratic part of the scatter matrix
S2 = D1 * D2; % combined part of the scatter matrix
S3 = D2 * D2; % linear part of the scatter matrix
T = - inv(S3) * S2; % for getting a2 from a1
M = S1 + S2 * T; % reduced scatter matrix
M = [M(3, :) ./ 2; - M(2, :); M(1, :) ./ 2]; %
premultiply by inv(C1)
[evec, eval] = eig(M); % solve eigensystem
cond = 4 * evec(1, :) .* evec(3, :) - evec(2, :) . 2;
% evaluate aCa
a1 = evec(:, find(cond > 0)); % eigenvector for min.
pos. eigenvalue
a = [a1; T * a1]; % ellipse coefficients
```

MATLAB code for an ellipse parametrization based on [11]

3   Hole Detection

After each shooting series, there should be no more than ten holes in the target. Each hole can be estimated by an ellipse and such ellipses can be detected using the Hough transformation [7].

The Hough transformation based ellipse center detection is a voting procedure. The point that gets the greatest number of votes wins. All results of Hough transformations can be ordered by the number of votes given to the center. Among them twenty best results are taken for future analysis.

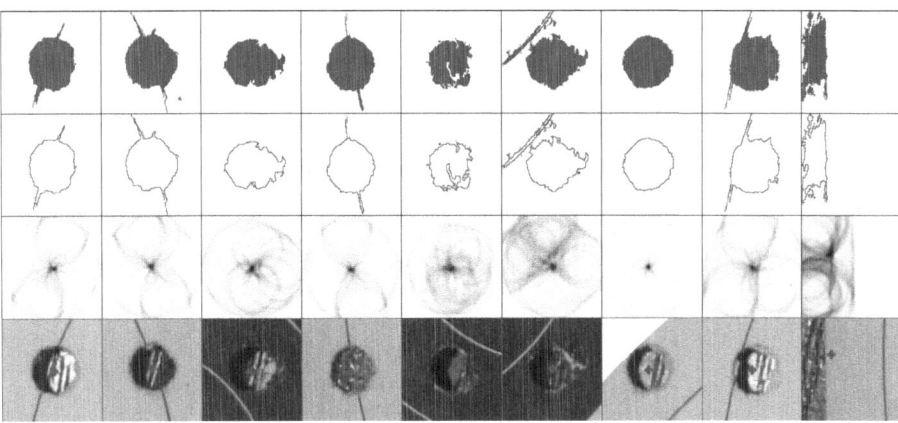

Fig. 5. Holes analysis. From the top: shapes created by the flood fill algorithm, detected edges, results of Hough transformation, and estimated centers of holes. The most right example is a false positive.

As an input for the Hough transformation results of the Prewitt operator can be used. The Prewitt operator is an edge detector that gave the best results among tested. However, rings, numbers, and other marks of the target have a significant influence on Hough transformation results. In Fig. 4 an input image 4(a) and Hough transformation results 4(b) are shown. Among the best results there are some false positives. Moreover, several existing holes are ignored because their results are placed outside the first twenty.

The results can be improved when detected rings (section 2.2) are erased from the image. In Fig. 4 an input image with erased rings 4(c) and Hough transformation results 4(d) are shown. This time a number of false positives is scarce and all holes are detected.

As alternative to Hough transformation, the median filter was also tested but used method worked faster.

4   Hole Analysis

The main aim of hole analysis is precise localization of hole center. Moreover, it can be used to reduce a number of false positive results.

The analysis of detected holes has the following steps. Firstly, the shape of hole is determined by the flood fill algorithm. If the algorithm is used on images different than binary then a color tolerance has to be defined. The distance between colors can be calculated as a maximum difference between components of the RGB color model:

$$d(c_1, c_2) = \max(|R(c_1) - R(c_2)|, |G(c_1) - G(c_2)|, |B(c_1) - B(c_2)|), \quad (7)$$

where $c_1, c2$ are colors and functions R, B, G calculate components of the RGB color model.

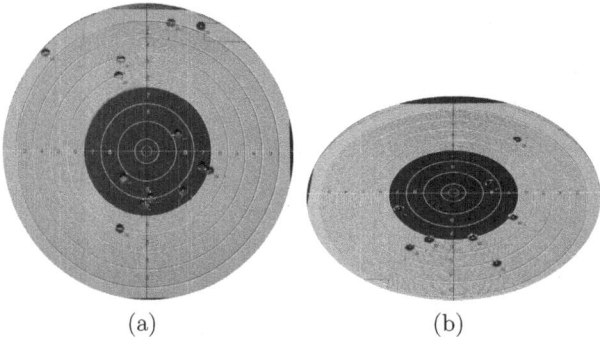

(a)   (b)

Fig. 6. The shooting targets. 6(a) Target with one hole missing. 6(b) Target with all holes detected on a photograph tilted by 30 degrees.

In the next step, the Prewitt operator is used to detect edges. The edges are then used as an input for the Hough transformation, which in turn is used to detect an ellipse center [4,5]. All steps are shown in Fig. 5.

Similarly as in the section 3 all results of Hough transformations can be ordered by the number of votes given to the center. The rejection threshold is calculated as 60 percent of the best result, which all results under the threshold being rejected. Remaining holes are classified as valid.

5   Results

In this section tests for genuine shooting targets are presented.

5.1   Number of Detected Holes

The tests were done for 14 shooting targets with 152 holes total. Pictures of the shooting targets were taken under various angles as it is shown in Fig. 6. The hole detection algorithm marked 208 objects. Among this number 150 were correctly detected holes and 58 were false positives. Details are given in Table 5.1. For each target the total number of holes (ALL) is shown as well as the number of detected ones (TP). The quality of detection is calculated as $\frac{TP}{ALL}$. The number of false positive (FP) decision is shown and its signification calculated as $\frac{FP}{FP+TP}$.

The detection results are very good. About 99 percent of holes were detected, which is better than the results published in [1]. However, among all positive decisions 28 percent were false positive.

The results of detection were used as input for the hole analysis described in Section 4. The analysis rejected all false positive inputs except two. The negative aspect is that 13 correct holes were also eliminated. Details are given in Table 2. For each target a number of positive (PI) and negative (NI) inputs from the detection is given. The quality of the analysis results is estimated by a classification error for each group of inputs.

Table 1. Results of holes detection

Shooting target	Holes	True positive	$\dfrac{TP}{ALL}$	False positive	$\dfrac{FP}{(FP+TP)}$
1	9	9	1.00	7	0.44
2	10	10	1.00	5	0.33
3	10	10	1.00	6	0.38
4	9	9	1.00	1	0.10
5	13	13	1.00	2	0.13
6	11	11	1.00	4	0.27
7	13	13	1.00	2	0.13
8	13	13	1.00	2	0.13
9	12	11	0.92	3	0.21
10	13	13	1.00	2	0.13
11	9	9	1.00	7	0.44
12	9	9	1.00	6	0.40
13	11	10	0.91	6	0.38
14	10	10	1.00	5	0.33
Total	152	150	0.99	58	0.28

The analysis rejected 97 percent of false positives inputs. However, about 8 percent of holes were rejected at the same time. The total quality of the algorithm calculated as a percent of detected holes multiplied by a percent of accepted in the analysis is 92 percent.

5.2   Quality of Estimated Score

When a center of a hole is known then its score can be calculated. A pellet diameter is constant and can be used to determine the radius of a hole. The scoring point lies on the circumference determined by both the hole's center and the pellet's radius. The point that lies nearest to the target center should be selected. The score is given by the smallest ring that includes the point.

This method of estimation was tested on 148 holes. The results were compared with human estimation . The distribution of errors is given in Fig. 7.

The average error is 0.05 points. For over 91 percent of holes, the error is less than 0.1 points, which is a typical scoring precision.

5.3   Comparison with Related Methods

There are two aspects of results evaluation. The first one is a number of detected holes as well as number of false positives. The second aspect is the score accuracy. Comparison of available papers papers concerning the subject is presented below.

In the paper [1] only holes detection is described. 98.3 percent of holes were detected. However, the number of false positives is not given. Similarly, our algorithm detects 99 percent of holes if false positives are not eliminated.

Table 2. Results of holes analysis

Shooting target	Input	Positive input	Negative input	False positive	$\frac{FP}{NI}$	False negative	$\frac{FN}{PI}$
1	16	9	7	1	0.14	1	0.11
2	15	10	5	0	0.00	1	0.10
3	16	10	6	0	0.00	1	0.10
4	10	9	1	0	0.00	0	0.00
5	15	13	2	0	0.00	1	0.08
6	15	11	4	0	0.00	1	0.09
7	15	13	2	0	0.00	1	0.08
8	15	13	2	0	0.00	0	0.00
9	14	11	3	0	0.00	0	0.00
10	15	13	2	0	0.00	1	0.08
11	16	9	7	0	0.00	1	0.11
12	15	9	6	0	0.00	0	0.00
13	16	10	6	0	0.00	1	0.10
14	15	10	5	1	0.20	1	0.10
Total	208	158	58	2	0.03	13	0.08

In the papers [8,12] only the accuracy of the total score is given. In both cases it is 0.1 point. It looks similar to our results (the average error is 0.05 points and the error is less than 0.1 for over 91)

In conclusion, our solution dedicated for amateur shooter gives similar results to the computer vision based systems for professional shooters.

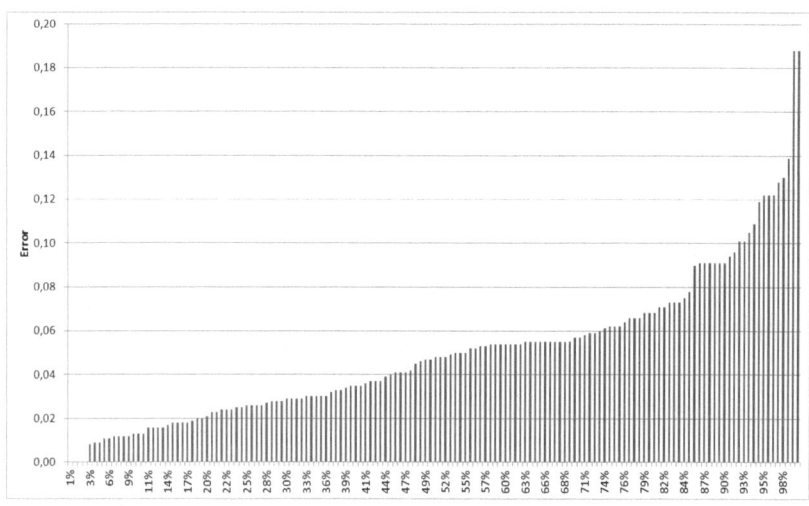

Fig. 7. Estimation errors distribution

6   Conclusions

In this paper, the issue of automatic scoring of shooting target is described. The proposed solution divides the issue into three problems: target detection, holes detection, and hole analysis.

The target detection based on a set of several basic algorithms is reliable. It localizes targets on amateur photos if only minimal requirements are satisfied.

The implemented hole detection finds 99 percent of holes in photos of at least 1 Mpix resolution.. Among detected holes false positives appear. Additional analysis eliminates most of them, with reduction of positive inputs to 92 percent.

The average error for the automatic score estimation is 0.05 points. For over 91 percent of holes the error is less than 0.1 points. The result is similar to the results of stationary systems for professional shooters.

The system is based on straightforward image processing such as the Prewitt edge detection and the Hough transformation. Simple techniques together with low requirements for processed images create possibility of developing the described system as an application for mobile devices.

References

1. Ali, F., Bin Mansoor, A.: Computer vision based automatic scoring of shooting targets. In: IEEE International INMIC 2008, pp. 515–519 (2008)
2. Aviatronic Ltd.: SmartSCORE Users Manual (2005),
   http://www.smartscore.hu/kep/users_manual_en.pdf
3. Sanctuary, C., Sean, A., Hsieh, S.R.: Remote strafe scoring system. United States Patent 4813877 (1989)
4. Chia, A., Leung, M., Eng, H.L., Rahardja, S.: Ellipse detection with hough transform in one dimensional parametric space. In: IEEE International Conference on Image Processing, ICIP 2007, vol. 5, pp. 333–336 (2007)
5. Davies, E., Barker, S.: An analysis of hole detection schemes, pp. 285–290 (1990)
6. DISAG-INTERNATIONAL: DISAG RM IV operating instruction (2005),
   http://www.disag.de/download/manuals/rmiv_en.pdf
7. Duda, R.O., Hart, P.E.: Use of the hough transformation to detect lines and curves in pictures. Commun. ACM 15(1), 11–15 (1972),
   http://doi.acm.org/10.1145/361237.361242
8. Fan, X., Cheng, Q., Ding, P., Zhang, X.: Design of automatic target-scoring system of shooting game based on computer vision. In: IEEE International Conference on Automation and Logistics, ICAL 2009, pp. 825–830 (August 2009)
9. Fichler, M., Bolles, R.: Random sample consensus: A paradigm for model fitting with applications to image analysis and automated cartography (1980)
10. Fitzgibbon, A.W., Pilu, M., Fisher, R.B.: Direct least-squares fitting of ellipses 21(5), 476–480 (May 1999)
11. Halir, R., Flusser, J.: Numerically Stable Direct Least Squares Fitting of Ellipses (1998), http://citeseerx.ist.psu.edu/viewdoc/summary?doi=10.1.1.1.7559
12. Liang, H., Kong, B.: A shooting training and instructing system based on image analysis, pp. 961–966 (August 2006)
13. SIUS ASCOR: Electronic Scoring systems (2010),
    http://www.sius.com/downloads/docu/Usermanual_System7_e.pdf

Using Bookmaker Odds to Predict the Final Result of Football Matches

Karol Odachowski and Jacek Grekow

Faculty of Computer Science, Bialystok University of Technology,
Wiejska 45A, Bialystok 15-351, Poland
odach@wp.pl, j.grekow@pb.edu.pl

Abstract. There are many online bookmakers that allow betting money in virtually every field of sports, from football to chess. The vast majority of online bookmakers operate based on standard principles and establish the odds for sporting events. These odds constantly change due to bets placed by gamblers. The amount of changes is associated with the amount of money bet on a given odd. The purpose of this paper was to investigate the possibility of predicting how upcoming football matches will end based on changes in bookmaker odds. A number of different classifiers that predict the final result of a football match were developed. The results obtained confirm that the knowledge of a group of people about football matches gathered in the form of bookmaker odds can be successfully used for predicting the final result.

Keywords: bookmaker odds, feature extraction, classification, forecasting, sports betting.

1 Introduction

The purpose of this paper is to investigate the possibility of predicting how upcoming sporting events will end based on changes in bookmaker odds. Football was the sport chosen for observation of changes in bookmaker odds. It should be assumed that if a gambler risks his own money, he has reasons to place such a bet. The greater the amount of gambled funds, the greater the change of the odds and greater possibility that the bet was based on factual knowledge about the competing teams, the status of the players, games played, etc. Predictions of the result can be based on such types of information. If the research should provide promising results, one might be tempted to build a decision-making system that could allow predicting final results based on observation of fluctuations of odds.

2 Previous Works

There are several papers that have dealt with similar problems of analysis and prediction of sporting event results. They are based on various types of data such as expert knowledge, results of previous matches, rankings of teams or bookmaker odds.

M. Graña et al. (Eds.): KES 2012, LNAI 7828, pp. 196–205, 2013.
© Springer-Verlag Berlin Heidelberg 2013

A group of papers directly referring to this paper are those addressing the problem of using data mining techniques to predict the final result of a sports match. An analysis of data of National Basketball Association (NBA) seasons was used to develop the expert system, which predicts the winner in a sport game [1]. The analyzed data contained detailed statistics of each game played during a season. The best accuracy (67%) was achieved by a classifier built using a multinomial logistic regression model with a ridge estimator. Miljkovic et al. [2] presents a system that uses data mining techniques in order to predict the outcomes of basketball games in the NBA league. To predict the game result the Naive Bayes method is used. Besides the actual result, the system calculates the spread for each game by using multivariate linear regression. Each game was described with attributes composed of the standard basketball statistics (field goals made, field goals attempted, 3 pointers, free throws, rebounds, blocked shots, fouls, etc), and information about league standings (number of wins and losses, home and away wins, current streak etc). The system correctly predicted the winners of about 67% of the matches. McCabe and Trevathan [3] used Artificial Neural Networks to predict games. They used attributes that indicate the quality of a particular team and achived 54.6% correct predictions for the English Football Premier League and 67.5% for Super Rugby. Smith et al. [4] used the Bayesian classifier to predict Cy Young Award winners in American baseball. The model was crated based on player statistics data collected for baseball seasons from 1967 to 2006. The accuracy of the Bayesian classifier was more than 80% correct.

3   Classic Football Bets of 1-X-2 Type

This work focuses on classic football bets of 1-X-2 type. For example, the result is a win of the first team, second team or a draw. Because of the three possible endings for the match, a three-way 1-X-2 bet, where "1" is an odd for the home team, "X" a draw, "2" odds for the away team, is used in football betting. The home team is the one that plays at home, while the visiting team is the away team. The classic bets are regarded as winning if the selected result is correct.

For example, a football match: Tottenham Hotspur vs. Chelsea:

- Bet on Tottenham Hotspur (type 1) will be settled as a win if Tottenham Hotspur wins. If Chelsea wins or there is a tie, the bets will be settled as a loss. It would be the same with a bet on Chelsea (type 2).
- A bet on a draw (type X) will be settled as a winning bet only in the event of the match ending in a tie.

4   Input Data

The input data describing the changes in bookmakers odds was obtained from the PinnacleSports [5] website, which makes public any information about sporting events in a clear form of an XML document. The XML file can be found at http://www.xml.pinnaclesports.com/pinnacleFeed.asp. This is a static file, updated every 10 minutes. The process of importing data from an XML file consisted in tracking its contents for the last 10 hours preceding the football game and recording data on the changing odds. Additionally, the input data had to be supplemented with the final result of the matches. Due to the fact that Pinnacle Sports does not provide such data, we imported it from another source: Betfair.com [6]. We collected the input data for a period of six months, from a total of 2615 matches.

4.1 Feature Extraction

Every game, which is an independent instance included in the input data of the decision-making system, was described by a set of features. They reflected significant changes in bookmaker odds, which may affect the final result of the match.

We analyzed the overall level of changes in bookmaker odds of football games, which could determine the path of further research. For this purpose sample graphs showing the odds over time were analyzed. A period of 10 hours of sampling before the match was taken into consideration, because in this period the greatest fluctuations of the odds occurred. The time interval between successive samples was 10 minutes.

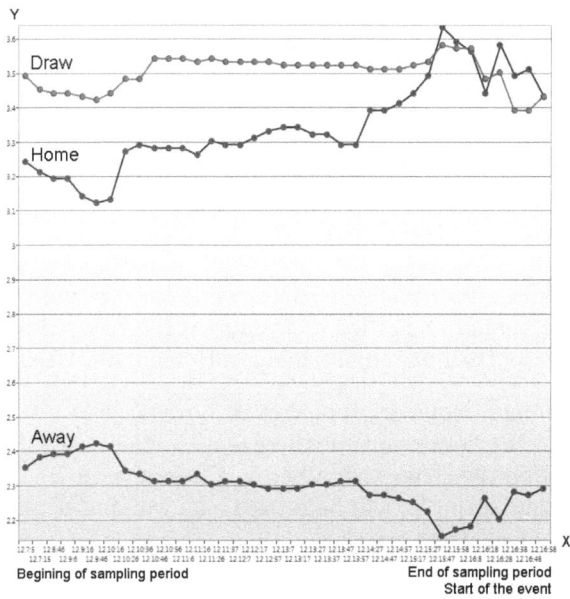

Fig. 1. Sample chart of odds changes (home, away, draw) in 1-X-2 type bets

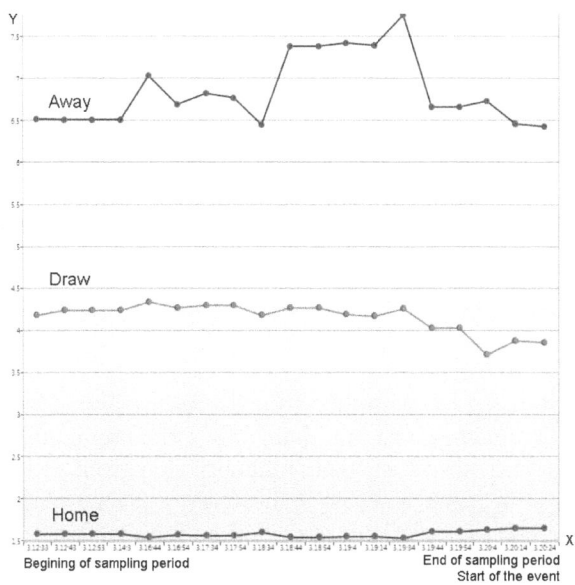

Fig. 2. Odd changes for the Racing Genk vs. Loceren (2:1) match

We observed that the closer to the start of the match, the more changes in the odds occurred. Figure 1 illustrates such a situation. This is a chart of values of the odds for the home team, the visiting team, and a draw over time (Y axis) during the last 10 hours (X axis) before the Tottenham Hotspur vs. Chelsea match, which was held on 12th December 2010 and ended with a 1-1 draw. Figure 2 presents another example of changes in bookmaker odds for the Racing Genk vs. Loceren (2:1) match, which was held on 3th April 2011.

We decided that it would be justified to divide the sampling period into several smaller ones, because the irregularity of the distribution of the changes may indicate that the entire sampling period does not have the same effect on the final result. For each period we generated the same set of features. Additionally, the entire sampling period was also taken into account. This allowed us to extract general information about the match. Figure 3 shows a schematic diagram of such a division.

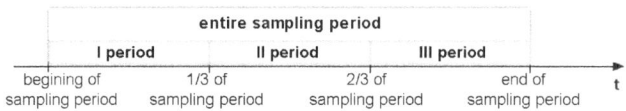

Fig. 3. Division of the sampling period

We examined three data sets:

1. Set of odds for a home team win;
2. Set of odds for a away team win;
3. Set of odds for a draw.

Regardless of the data set, we investigated four sampling periods:

- the entire 10-hour sampling period prior to the match;
- 1st sampling period correlating to the first 3 hours and 20 minutes;
- 2nd sampling period beginning on the 21st minute of the third hour and also lasting 3 hours and 20 minutes;
- 3rd sampling period correlating to the last 3 hours and 20 minutes before the start of the match.

For each set of data in each sampling period we generated a set of 24 *standard features*, which include: the minimum value; the minimum value given as a percentage; the maximum value; the maximum value given as a percentage; the value of the arithmetic mean; arithmetic mean given as a percentage; the number of odds with different values; standard deviation; initial value; initial value given as a percentage; final value; final value given as a percentage; the difference in the initial and final value; the difference in the initial and final values given as a percentage; angle between the horizontal line and a line drawn from the initial to the maximum value; angle between the horizontal line and a line drawn from the initial to the minimum value; minimum value of the derivative; maximum value of the derivative; arithmetic mean of the derivative; standard deviation of the derivative; initial value of the derivative; final value of the derivative; the difference between the initial and the final values of the derivative; the number of different values of the derivatives.

Additionally, a single sampling period contained eight *general features* that apply to all three data sets simultaneously: minimum limit of money; maximum limit of money; arithmetic mean of the limit of money; nominal feature which based on the arithmetic mean value of the odds determines the favored team; nominal feature determining the favored team at the beginning of the sampling period; nominal feature determining the favored team at the end of the sampling period; nominal feature determining the team that recorded the biggest odds drop between the beginning and the end of the sampling period; nominal feature determining the team that recorded the biggest odds drop between successive sampling periods.

The number of features determined for a single sampling period is equal to 80. This is the sum of the *general features* (8) and the product of the number of features included in the set of *standard features* (24) with the amount of data sets (3). We get a total of 320 features from the four sampling periods.

4.2   Data in ARFF Format

To use the collected input data about the matches in the decision-making process, the values of all the features describing a particular match were determined and later recorded in the ARFF file format. The last declared attribute (feature) in this file is the decision class, which is the result of the match and adopts the nominal values from the set: Win-home, Win-away, Win-draw. It defines the final outcome of the match. For the input data prepared in such a manner, classifiers were developed allowing to predict the final result. To analyze the data and the development of classifiers, a data mining task software WEKA [7] was used. Cross-Validation Folds 10 (CV-10) were used to evaluate the classifiers.

5   Experiment Results

We constructed three variants of classifiers in order to thoroughly test the data on football matches.

5.1   Standard Data Set Classification

To make the data collected from the PinnacleSports and Betfair sites useful for data mining purposes, they had to go through pre-treatment in the form of transformation and cleaning of the collected information. The overall objective was to minimize so-called GIGO (garbage in - garbage out) - the reduction of "garbage" that enters the model so that the model could minimize the number of incorrect results [8]. For this purpose, the study included only those events that had odds in the full 10-hour sampling period and had not been postponed. An equal number of matches for each decision-making class was included in order to offset the number of instances from each class [9]. Thus a total of 1116 sample football games were selected, including: 372 matches that ended with a win for the home team; 372 matches that ended with a win for the away team; 372 matches that ended with a draw.

Six classification algorithms were selected: BayesNet, SMO, LWL, EnsembleSelection, DecisionTable and SimpleCart [7]. For attribute selection the following attribute evaluators and search methods were used: CfsSubsetEval with BestFirst, CfsSubsetEval with LinearForwardSelection and PrincipalComponents with Ranker. The highest accuracy rate of 46.51% was achieved by the DecisionTable algorithm. The confusion matrix for the created model is presented in Table 1.

Table 1. Confusion matrix of classifier for a win for the home team, the away team or a draw

a	b	c	← classified as
260	65	47	a = Win-home
154	154	64	b = Win-away
173	94	105	c = Win-draw

Matches that ended with a win for home team (Win-home class) are classified very well in comparison with the two other classes. Most of the matches which ended with a win for the away team were classified slightly worse. In this case a big mistake occurred due to a mistaken classification as a win for the home team. The worst is the classification of matches that ended in a draw, which are mostly classified incorrectly as a win for the home or the away team. This is because a draw is a middle class between the two results.

5.2 Binary Classification

For better detection of the match result, we decided to build binary classifiers [10] for each type of match result: win for the home team (Win-home), win for the away team (Win-away), and a draw (Win-draw). The binary classifier focuses on one problem and it can perform a better classification than a classifier that has to identify three classes.

Binary Classifier for a Win for the Home Team.   When developing a classifier for the home team win, just as before (section 4.1), we used 1116 sample football matches. Matches which ended with a win for the home team remained unchanged, but the matches that ended with a win for the visiting team and a draw were combined to form a new class. Then, we randomly discarded 372 matches to make the number of the instances in each class equal. Below is the size of the two classes: 372 matches that ended with a win for the home team (Win-home class); 372 matches that ended with a win for the away team or a draw (Win-no-home class).

Six classification algorithms were selected: BayesNet, SMO, LWL, Bagging, DecisionTable, and LadTree. For attribute selection the following attribute evaluators and search methods were used: CfsSubsetEval with BestFirst, ConsistencySubsetEval with GreedyStepwise, WrapperSubsetEval (classifier: Bagging) with BestFirst. The highest accuracy rate of 70.56% was noted by the Bagging algorithm, which obtained this result after feature selection (WrapperSubsetEval with BestFirst) and after discretization of attributes. The confusion matrix for the created model is presented in Table 2.

Table 2. Confusion matrix of binary classifier for a win for the home team

a	b	← classified as
229	143	a = Win-home
76	296	b = Win-no-home

Binary Classifier for a Win for the Away Team. The accuracy of predicting a win for the away team proved to be a bit more difficult than predicting a win for the home team. The classifiers achieved worse results, but as in previous studies, a positive influence of feature selection and data discretization was observed. The highest accuracy rate of 65.46% was noted by the Bayesian NaiveBayes algorithm. The confusion matrix for the created model is presented in Table 3.

Table 3. Confusion matrix of binary classifier for a win for the away team

a	b	← classified as
244	128	a = Win-no-away
129	243	b = Win-away

Binary Classifier for a Draw. Same as with the evaluation of classifiers of the standard data set (section 4.1), draws proved to be very difficult to predict. In many cases, the classifiers could not perform a correct classification, which resulted in obtaining accuracies which were not satisfactory. It can be concluded that the values of features describing matches that ended in a draw are very similar to those relating to the win of the home and away teams. The Ensemble-Selection classifier proved to be the most accurate which after feature selection (without discretization) achieved an accuracy of 56.99%. The confusion matrix for the created model is presented in Table 4.

Table 4. Confusion matrix of binary classifier for a draw

a	b	← classified as
196	176	a = Win-no-draw
144	228	b = Win-draw

5.3 Classification of Data without Draws

Due to the fact that predicting a draw is difficult, we decided to perform additional tests on data that do not contain instances of matches ending in a draw. This allowed creating a classifier that could enable predicting a win for the home or the away team. This information can be used to place *Asian handicap* bets, where in the case of a draw the betting amount is returned.

Matches that ended in a draw were discarded from the 1116 football matches sample set. Matches that ended with a win for the home or the away team were left unchanged. Below is the size of the two classes: 372 matches that ended with a win for the home team (Win-home class); 372 matches that ended with a win for the away team (Win-away class).

Six classification algorithms were selected: BayesNet, VotedPerception, Ibk, Bagging, DecisionTable, and LADTree. For attribute selection the following attribute evaluators and search methods were used: CfsSubsetEval with BestFirst, ConsistencySubsetEval with BestFirst, WrapperSubsetEval (classifier: Naive-Bayes) with BestFirst.

Removal of matches that ended in a draw from the sample data set proved to be very beneficial. Classifiers predicting a win for a home or away team obtained the highest accuracy taking all the conducted studies into account. The classifier that proved to be the most accurate was an algorithm based on the Bayesian network: BayesNet, which after feature selection conducted after discretization achieved an accurancy of 70.30%. The confusion matrix for the created model is presented in Table 5. The best BayesNet algorithm correctly classified more than 80% of Win-home and 60% of Win-away class matches.

Table 5. Confusion matrix of classifier for win for the home or the away team

a	b	← classified as
298	74	a = Win-home
147	225	b = Win-away

5.4   Summary of Classification of 1-X-2 Type Bets

The evaluation performed on the classifiers built for 1-X-2 type bets showed that a draw is the most difficult to predict. This study confirms the reality of football, because the draw class determines the intermediate odd between a win for the home and the away team. Tests showed that features describing a draw contain many similarities to those relating to a win for the home or the away team. Matrices of classification errors in the study of the standard data set show that most matches which ended in a draw are incorrectly classified as a win for the home team. This is due to the fact that in most cases, the home team is the favorite (has the lowest odd).

In the case of binary classifiers, the accuracy of predicting a win for the home team and the away team is promising. The classifier of a win for the home team achieved an accuracy of 70.56%. Once again the classifier of a draw had the worst results. The best independent classifier was the classifier of a win for the home or away team; the accuracy did not deteriorate with matches which ended in a hardly recognizable draw. The achieved accuracy of this classifier is very satisfying. This classifier can be used for *Asian handicap* bets, where in the case of a draw the betting amount is returned.

In most cases, feature selection resulted in increasing the accuracy of classification. We observed that the features were selected from all the sampling intervals. A selection frequently used features concerning the *minimum* and *maximum values, angles to these values, derivatives*, the *differences* between the first and last samples in the interval, and the *largest drops* in the value of odds between adjacent samples. This indicates that these features were most important.

Table 6. Classifying algorithms selected for predicting 1-X-2 type bets

Type of classifier	Algorithm	Accuracy
Standard data set	DecisionTable	46.51%
Win for home team	Bagging	70.56%
Win for away team	NaiveBayes	65.46%
Draw	EnsembleSelection	56.99%
Win for home and away team	BayesNet	70.30%

Discretization in most cases also had a very positive influence on the results of classification. Below are the best classification algorithms that have been selected to predict the final results of new football matches. A summary of accuracy of the developed classifiers is presented in Table 6.

6 Conclusions

The results obtained, an effectiveness of 70%, are quite satisfactory and prove the existence of a relationship between changes in the bookmaker odds values and the outcome of the football match. These results confirm that the knowledge of a group of people about football matches gathered in the form of bookmaker odds can be successfully used for predicting the final result. Based on our research results, one could build a decision-making system that could allow predicting final results based on observation of fluctuations of odds. In further work on the system, new features describing changes of the odds should be investigated, which would probably contribute to improving the accuracy of the system.

Acknowledgments. This paper is supported by the S/WI/5/08.

References

1. Zdravevski, E., Kulakov, A.: System for Prediction of the Winner in a Sports Game. In: ICT Innovations 2009, Part 2, pp. 55–63 (2010)
2. Miljkovic, D., Gajic, L., Kovacevic, A., Konjovic, Z.: The use of data mining for basketball matches outcomes prediction. In: 8th International Symposium on Intelligent Systems and Informatics, pp. 309–312 (2010)
3. McCabe, A., Trevathan, J.: Artificial Intelligence in Sports Prediction. In: Proceedings of the Fifth International Conference on Information Technology: New Generations, pp. 1194–1197. IEEE Computer Society (2008)
4. Smith, L., Lipscomb, B., Simkins, A.: Data Mining in Sports: Predicting Cy Young Award Winners. Journal of Computing Sciences in Colleges, Consortium for Computing Sciences in Colleges 22(4), 115–121 (2007)
5. Pinnacle Sports, http://www.pinnaclesports.com
6. Betfair, http://www.betfair.com
7. Hall, M., Frank, E., Holmes, G., Pfahringer, B., Reutemann, P., Witten, I.H.: The WEKA Data Mining Software: An Update. SIGKDD Explorations 11(1) (2009)
8. Larose, D.T.: Discovering Knowledge in Data: An Introduction to Data Mining, p. 28. Wiley Interscience (2005)
9. Hand, D., Mannila, H., Smyth, P.: Principles of Data Mining. MIT Press (2001)
10. Witten, I.H., Frank, E.: Data Mining: Practical machine learning tools and techniques. Morgan Kaufmann, San Francisco (2005)